# World Symposium on Smart Materials and Applications (WSSMA 2021)



IOP Conference Series: Materials Science and Engineering
Volume 1186

Bangkok, Thailand
16-18 July 2021

ISBN: 978-1-7138-4646-8
ISSN: 1757-8981

**Printed from e-media with permission by:**

Curran Associates, Inc.
57 Morehouse Lane
Red Hook, NY 12571



Institute of Physics
Dirac House, Temple Back
Bristol BS1 6BE UK

Phone: 44 1 17 929 7481
Fax: 44 1 17 920 0979

techtracking@iop.org

**Additional copies of this publication are available from:**

Curran Associates, Inc.
57 Morehouse Lane
Red Hook, NY 12571 USA
Phone: 845-758-0400
Fax: 845-758-2633
Email: curran@proceedings.com
Web: www.proceedings.com

# TABLE OF CONTENTS

IOP Conf. Series: Materials Science and Engineering 1186 (2021) 011001 doi:10.1088/1757-899X/1186/1/011001

# *Preface*

The World Symposium on Smart Materials and Applications (WSSMA 2021) was taken place in Bangkok, Thailand during July 16-18, 2021.

This conference brought together leaders from industry and academia to exchange and share their experiences, present research results, explore collaborations and to spark new ideas, with the aim of developing new projects and exploiting new technology in the field of applied materials and advanced materials processing technologies.

The committee of WSSMA 2021 expresses their sincere thanks to all authors for their high-quality research papers and careful presentations. All reviewers are also thanked for their careful comments and advices. Thanks, are finally given to Trans Tech Publication as well for producing this volume.

The Organizing Committee of WSSMA 2021

Committee Chair

Prof. Jongwan Hu | Incheon National University

Incheon Disaster Prevention Research Center

IOP Conf. Series: Materials Science and Engineering 1186 (2021) 011001 doi:10.1088/1757-899X/1186/1/011001

## *Conference Photograph*

IOP Conf. Series: Materials Science and Engineering 1186 (2021) 011001 doi:10.1088/1757-899X/1186/1/011001

***Sponsors***

| | |
|---|---|
| Incheon Disaster Prevention Research Center (IDPRC) | I D P R C 인천대학교 방재연구센터 Incheon Disaster Prevention Center |

Recently, various efforts to prevent and prepare are vitally needed for prevention of Disasters and calamities. So we understand the necessity for technology of disaster and we built up the Incheon Disaster Prevention Research Center (IDPRC) in Incheon National University (INU) in 1997.

Accordingly, Incheon Disaster Prevention Research Center (IDPRC) in Incheon National University has progressed of research on the prevention of disasters and calamities through the various seminars, conference and lectures. This research could be conducted cooperation with Incheon National University (INU) in various fields' structure, soil, hydraulic and environment.

Incheon Disaster Prevention Research Center (IDPRC) will try to be a leader in the disaster of industry through the various research activities and global conference.

IOP Conf. Series: Materials Science and Engineering 1186 (2021) 011001 doi:10.1088/1757-899X/1186/1/011001

## *Lists of Committees*

| Committee name | Organizer name | Organization |
|---|---|---|
| Honorary Chair | Jongwan Hu | Incheon National University |
| Conference General Chairs | Mosbeh Kaloop | Incheon National University |
| Honorary Chair | Dongkeon Kim | Dong-A University |
| Technical Committees | J. W. Hu | Incheon National University |

IOP Conf. Series: Materials Science and Engineering 1186 (2021) 011002 doi:10.1088/1757-899X/1186/1/011002

# Peer review declaration

All papers published in this volume of *IOP Conference Series: Materials Science and Engineering* have been peer reviewed through processes administered by the Editors. Reviews were conducted by expert referees to the professional and scientific standards expected of a proceedings journal published by IOP Publishing.

- **Type of peer review:**
  All papers revise by three of committee members of the conference. The papers are divided based on the specialist of the committee members; and in the case of special topics, other reviewers invite for reviewing the selected manuscripts.

  Also, reviewers have considered the Technical Criteria (Scientific merit: notably scientific rigour, accuracy and correctness. Clarity of expression; communication of ideas; readability and discussion of concepts. Sufficient discussion of the context of the work, and suitable referencing) and Quality Criteria (Originality: the work relevant and novel, Motivation: Does the problem considered have a sound motivation? All papers should clearly demonstrate the scientific interest of the results. Repetition: Have significant parts of the manuscript already been published? Length: Is the content of the work of sufficient scientific interest to justify its length?)

- **Conference submission management system:**
  Our conference has Secretary for management on conference papers. So, authors send their papers to Conference E-mail, and our secretary has managed it. The secretary sends the papers to the conference chairs to evaluate the suitability of submitted papers to the conference scopes.

- **Number of submissions received:**
  Especially in this time, only few of papers were submitted cause of COVID-19. The total number of submission papers were 34 papers.

- **Number of submissions sent for review:**
  The submission papers were revised from the management conference system and committee organizations and the papers that sent for review were 25 papers.

- **Number of submissions accepted:**
  The papers that accepted in the conference for publishing after review processing are 8 papers.

- **Acceptance Rate (Number of Submissions Accepted / Number of Submissions Received X 100):** 23 %

- **Average number of reviews per paper:**
  Different cycles review processing implemented for improving the papers quality. In average, three times review on per paper were processed.

- **Total number of reviewers involved:**

IOP Conf. Series: Materials Science and Engineering 1186 (2021) 011002 doi:10.1088/1757-899X/1186/1/011002

Three of reviewers are took part in the review process. When we receive positive reports from two at least we can accept the paper after evaluating the reviewer report and authors responses.

- **Any additional info on review process:**
  The processing of review work of this conference was carried out as follows: (1) the secretary received the papers by emails and processed the initial checking of the papers, he revised the format, language and roughly check of papers contents; (2) the general chairs received the processed papers and they revised the papers scopes, Technical Criteria (Scientific merit: notably scientific rigour, accuracy and correctness. Clarity of expression; communication of ideas; readability and discussion of concepts. Sufficient discussion of the context of the work, and suitable referencing) and Quality Criteria (Originality: the work relevant and novel, Motivation: Does the problem considered have a sound motivation? All papers should clearly demonstrate the scientific interest of the results. Repetition: Have significant parts of the manuscript already been published? Length: Is the content of the work of sufficient scientific interest to justify its length?)
  If these papers relate to the conference scopes it can be passed to the next processing, else the papers can be rejected. (3) all accepted papers divided into categories based on the papers contents and sent to three committee members who have a good experience in the paper contents. (4) after receiving the review reports from the committee members, the review reports sent to authors for sending their responses on the listed comments. (5) the reviewing cycling process divided into accept, minor and major revisions; and based on the review comments and suggestion, the papers can be processing up to three times to improve the manuscript contents. (6) finally, the authors should revise/argue the comments and send their response through email; and the processing cycling until organizer accept the paper according to the final review reports.

- **Contact person for queries:**
  Jack Son (E-mail: wssma2021@wssma-conf.org)

# Application of DLVO Modeling to Study the Effect of Silica Nanofluid to Reduce Critical Salt Concentration in Sandstones

**Rizwan Muneer[1,a*], Muhammad Rehan Hashmet[2,b] and Peyman Pourafshary[3,c]**

[1,2,3]School of Mining and Geosciences, Nazarbayev University, Nur-Sultan 010000 Kazakhstan

[a]rizwan.muneer@nu.edu.kz, [b]muhammad.hashmet@nu.edu.kz, [c]peyman.pourafshary@nu.edu.kz

**Abstract.** Critical salt concentration (CSC) is the minimum salt concentration of injected water used for different oil recovery operations, below which fines migration initiates within sandstone reservoirs having clay contents and could be one of the potential causes of formation damage. This paper estimated CSC for Berea sandstone-NaCl brine system experimentally and by using DLVO modeling based on Van der Waals and electric double layer (EDL) surface forces. Furthermore, the effectiveness of silica nanofluid to reduce CSC and control fines migration was investigated. At different salinities of the injected fluid, the experimental CSC was determined by performing corefloods and analyzing effluent turbidity and absorbance. Attractive (van der Waals) and repulsive (EDL) forces were estimated by DLVO theory to predict the value of CSC. The experimental data were used to validate the model. The application of silica nanoparticles (NPs) was studied by injecting 0.1 wt% silica nanofluid and the resultant reduction in CSC was observed. During the injection of brines at different salinities ranging from 0.2 M to 0.05 M into the sandstone porous media, fines production was detected in the effluent after the injection of 0.1M NaCl solution, indicating that the CSC was 0.1M for this sand-fine-brine (SFB) system. The zeta potentials for SFB systems of different salinities were measured by zeta-sizer and were in the range of -32 mV to -24 mV. DLVO model based on attractive and repulsive forces was then used to predict the CSC, at which the total DLVO energy shifts from negative to positive and was around 0.1M for NaCl solution. The model prediction was in close agreement with the experimental results. Both the experiment and DLVO model showed the efficiency of silica nanofluid to mitigate fines migration as no fines production was observed for 0.1M NaCl case after the treatment with nanofluid. The estimation of CSC is critical to avoid formation damage during many oil recovery processes, included but not limited to low salinity waterflooding and alkaline flooding. The utilization of nanofluid provided promising results in controlling fines migration and reducing CSC. The proposed model based on DLVO theory can work as a useful tool to predict CSC without the need for extensive experimental work.

## 1. Introduction

Oil-containing sandstone reservoirs are a huge source of energy worldwide and consist of about 60% of the world's petroleum reservoirs [1]. Due to low salinity waterflooding, a decrease in ionic strength and change in ionic composition of formation water with different cations and anions can cause fines migration and permeability impairment in porous and permeable sandstones containing different types of clay minerals. These minerals are a part of rock bulk volume and reside in the pores either in the form of clusters or as fines covering the sand grains. Clay minerals are primarily composed of aluminosilicates with a platelet configuration. The chemical composition of clay minerals incorporates silicon oxide, aluminum oxide, and compounds of magnesium. Kaolinite, illite, and montmorillonite are the three most commonly found clay minerals [2]. The fine particles remain in equilibrium in a high salinity formation water environment as the electric double layers around fine and sand grain are contracted because of the presence of high charge density in the system. However, when system salinity decreases due to the injection of low salinity water, both electric double layers expand, and because of high repulsion, fine particles are released and migration starts within the reservoir even at a low flow rate [3]. Released fines can block the pore throats and cause productivity and injectivity problems due to high pressure drop requirements [4, 5]. Fines can be released because of three conditions: (1) if the system salinity is less than a critical salt concentration (2) production or injection velocity is higher than a critical velocity and, (3) pH of the system is greater than a critical pH [6].

The use of nanotechnology to control the fines migration has gained considerable importance in the petroleum industry in recent years. NPs are very small particles with a size ranging between 1–100 nanometers and having a high specific surface area due to their extremely small size. The adsorption of NPs onto the rock surface changes the surface potential. Various types of NPs with different chemical nature (composition, water solubility, and stability) and distinctive properties (size, shape, and specific surface area) have been utilized to mitigate fines migration e.g. silicon oxide ($SiO_2$) [7, 8], magnesium oxide (MgO) [9], and aluminum oxide ($Al_2O_3$) [10]. Derjaguin, Landau, Verwey, and Overbeek (DLVO) presented a theory that is a useful tool to model the dynamics of a sand-fine-brine system and study the effect of various critical parameters on fine migration initiation [11-13]. It describes the total potential ($V_T$) of the system by integrating all the effective surface forces which include van der Waals attractive potential ($V_V$), electric double layer potential ($V_E$), and Born repulsion ($V_B$). This theory considers that the $V_V$, $V_E$, and $V_B$ do not depend on each other, and hence can be summed up for the particle-plate configuration, to quantify the total potential as a function of separation distance. The DLVO-based total potential of a system containing a fine particle and a pore surface is presented as Equation-1 which is a superposition of all the energies. In this research work, the application of nanoparticles to reduce CSC was examined experimentally and modeled by the analysis of surface forces using DLVO theory.

$$V_T = V_V + V_E + V_B. \quad (1)$$

## 2. Methodology

### *2.1. Brines.*

To replicate the reservoir conditions, high salinity formation water (FW) of 7.7 wt% ionic strength was used to saturate the core samples in the experimental phase. NaCl brines of 0.2, 0.15, and 0.1 molarities with pH 7-7.5 were used to determine the CSC for the given SFB system. The ionic compositions of the brines are given in Table 1.

Table 1. Composition of formation water and NaCl brines.

| Ions | FW | 0.2M NaCl | 0.15M NaCl | 0.1M NaCl |
|---|---|---|---|---|
| | [ppm] | | | |
| $Na^+$ | 23426 | 4600 | 3450 | 2300 |
| $Ca^{2+}$ | 4448 | - | - | - |
| $Mg^{2+}$ | 1300 | - | - | - |
| $Cl^-$ | 47781 | 7093 | 5320 | 3547 |
| Total | 76955 | 11693 | 8770 | 5847 |

### *2.2. Rock Samples.*

Berea sandstone outcrop cores were used for the experimental part. The mineral composition of the Berea sandstone as obtained from scanning electron microscopy - energy dispersive spectroscopy

(SEM-EDS), and was found to be 46.2 wt% Si, 35.7 wt% O, 2.1 wt% Al, 1.1 wt% K, and 0.35 wt% Fe. Four core samples were cut, each with 3 in length and 1.5 in dia. The samples were initially oven dried and then saturated with formation water using Vinci manual saturator. The wet weights were measured and porosities for each sample were calculated. Each sample was then flooded with FW using the Vinci SRP30 coreflood system to measure absolute permeabilities. Table 2 shows the properties of each sample. The core samples were then used for different experiments.

Table 2. Properties of sandstone cores.

| Sample ID | Diameter | Length | Pore volume | Porosity | Permeability |
|---|---|---|---|---|---|
| | [mm] | | [cc] | [%] | [md] |
| 1 | 38.12 | 75.98 | 17.40 | 19.21 | 295.3 |
| 2 | 38.12 | 76.45 | 16.77 | 19.22 | 303.5 |
| 3 | 38.12 | 76.20 | 17.05 | 19.65 | 320.1 |
| 4 | 38.12 | 74.70 | 16.65 | 19.53 | 312.1 |

*2.3. Nanofluid.*
To study the effect of nanoparticles on CSC, silica nanoparticles of 20 nm size were used provided in the form of a 25 wt% concentrated dispersion by Glantero Ltd. Ireland. The reason for selecting these nanoparticles is their hydrophilic nature, high stability, low toxicity, and better performance in controlling fines migration in sandstones [14]. Moreover, 0.1 wt% nanofluid was prepared by diluting the dispersion using 0.15M NaCl brine.

*2.4. Coreflooding.*
Three corefloods were performed by injecting brines of different molarities at a flow rate of 0.2 cc/min (0.82 ft/day) and pressure drop was monitored to estimate the fines migration initiation. Effluent absorbance was measured by UV-Vis-Spectrophotometer and turbidity was obtained by turbidity meter to confirm fines presence and determine CSC for NaCl solution. A fourth coreflood was conducted by injecting 0.1 wt% silica nanofluid at the observed CSC to analyze the effectiveness of NPs in controlling fines migration and reducing the CSC.

*2.5. Zeta Potential Measurements.*
Sand-brine and sand-nanofluid zeta potentials were required as input for DLVO modeling. For this purpose, a core sample was crushed using a cone crusher and sieved to 40 um size of sand grains. The sieved sample was washed with a sequence of distilled water, acetone and again distilled water to remove any impurities on the sand grains. The crushed sand was dispersed in different brines and nanofluid with an ultrasonic homogenizer and the zeta potentials were measured using Malvern zeta sizer.

*2.6. Fines Analysis.*
The effluents from coreflood tests were oven-dried to collect the in-situ fines and the shape and average size of a fine particle were obtained using SEM. Furthermore, SEM-EDS was performed to obtain information regarding the chemical composition of fines.

*2.7. DLVO Modeling.*
Measured zeta potentials, the average size of fine particles, and ionic strength of the brines were used as input for DLVO modeling, and surface forces were calculated based on available models mentioned in our previous study [15]. CSC was determined using the DLVO model and was compared with the experimentally obtained CSC to validate the proposed approach. In addition, the model was used to check the efficacy of nanofluid in altering the surface forces and minimizing the fines movement.

IOP Conf. Series: Materials Science and Engineering 1186 (2021) 012001 doi:10.1088/1757-899X/1186/1/012001

## 3. Results and Discussion

### *3.1. CSC from Corefloods.*

The first objective of coreflood tests was to determine the CSC for NaCl brine-sandstone system. The pressure drop versus pore volume data was recorded for each case and a high pressure drop was observed during the 0.1M NaCl injection scenario as compared to 0.2M and 0.15M brines injection. This high pressure drop was the indication of fines release and migration in the system at 0.1M ionic strength. However, the other two scenarios showed a relatively low and stable pressure drop during the experiment, indicating 0.1M was the CSC for the SFB system under study. Moreover, turbidity analysis (Fig. 1a) and UV-spectrophotometry results (Fig. 1b) also showed that the effluent produced during 0.1M NaCl injection had high turbidity and more light absorbance due to the presence of a high concentration of fine particles. Effluent received after 0.1M NaCl injection had high turbidity of 25.6 NTU whereas 0.15M and 0.2M effluents showed turbidities of 2.51 and 1.71 NTU respectively, which could be because of some mineral dissolution in the cores. Similarly, the same effluent of 0.1M injection resulted in light absorbance of 0.458 ABS, and the other two effluents showed negligible light absorbance of 0.05 and 0.019 ABS. These experimental results collectively confirmed the fines migration after 0.1M brine injection. Hence, the CSC from the experimental results was around 0.1M.

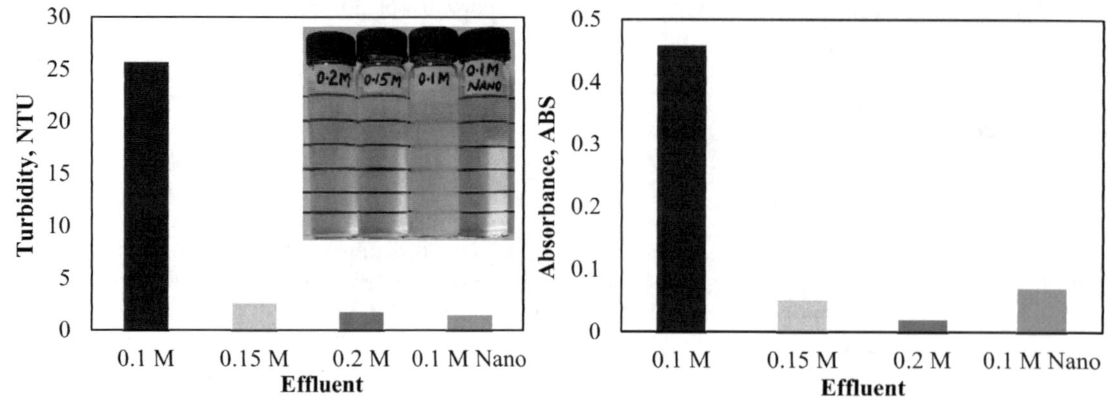


(a) (b)

**Figure 1.** Results of effluents (a) Turbidity, and (b) Absorbance as a function of ionic strength.

### *3.2. Fines Characterization.*

The fines recovered from the coreflood effluent were analyzed by SEM and SEM-EDS and the results are presented in Figure 2.

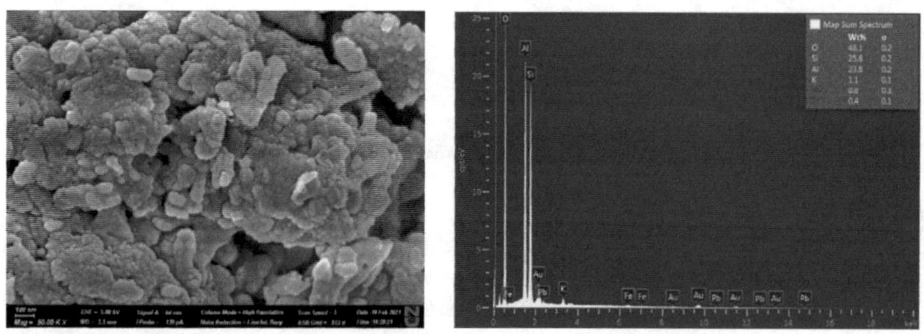

(a) (b)

**Figure 2.** (a) SEM image of kaolinite fines, and (b) SEM-EDS results for fines composition.

Based on the SEM-EDS and X-ray diffraction (XRD) analysis, the fine particles were composed of mainly kaolinite clay ($Al_2Si_2O_5(OH)_4$), and the average size of these particles was around 800-900 nm.

### *3.3. CSC by DLVO Modeling.*

Zeta potentials were measured for different systems being studied and were found to be in the range of -32 mV to -24 mV for the brine salinity range of 0.1M to 0.2M (Fig. 3a). The measured zeta potentials were in close agreement with the developed correlations in our previous study [15]. An increase in zeta potential towards the negative side was observed with decreasing brine salinity, indicating the dominance of repulsive forces in the system. The measured zeta potentials, ionic strength of each brine, and the average size of fine particles were used to calculate surface forces between fine and sand grain, and then the total interaction energy of each system was calculated using DLVO modeling as shown in Figure 3b. The total interaction energy is the sum of attractive and repulsive forces. In the high salinity environment, as in the case of 0.2M NaCl solution, attractive forces dominate the repulsive forces because of compacted electric double layers on fines and sand grains and the total energy comes out to be negative, which indicates there would not be any fines migration in the system. The same results were obtained for 0.2M, and 0.15M brine environments with the total interaction energy in the negative region. However, for 0.1M brine, the total interaction energy shifted from negative to positive indicating the dominance of repulsive forces in the system at this separation distance. This can be attributed to the expansion of electric double layers around fines and sand grains under low salinity conditions. Based on the DLVO model results, it was inferred that 0.1M was less than CSC, and CSC was in between 0.15M and 0.1M NaCl brines for the SFB system under investigation. Thus, the DLVO model validated the experimental results and can be successfully used to predict CSC for a certain brine-rock system.

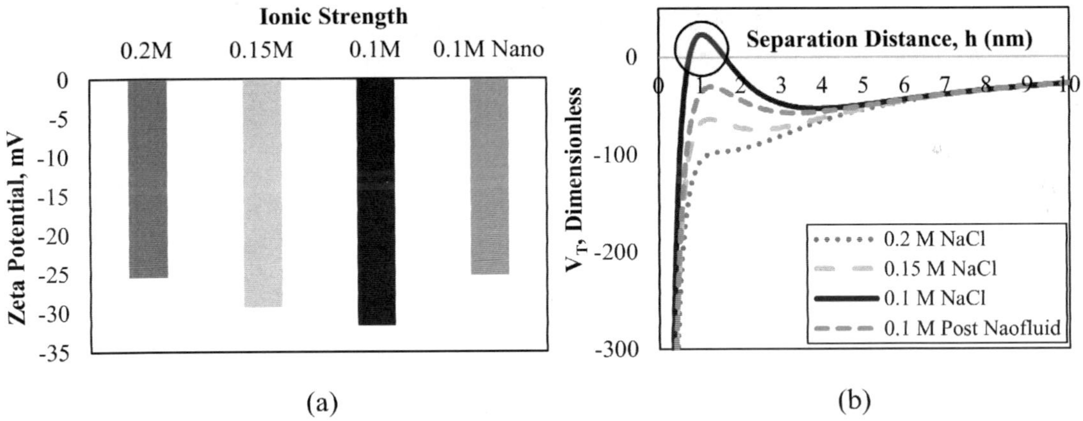


**Figure 3.** (a) Zeta potentials for different systems, and (b) CSC prediction from DLVO model.

### *3.4. Effect of Nanoparticles on CSC.*

In the last coreflood test, 0.1 wt% silica nanofluid was injected in the core at an injection rate of 0.02 cc/min after FW injection. This low injection rate was set to provide more time for the NPs to interact with the sandstone surface and alter the surface forces. After injecting 4 PVs of nanofluid, 24 hrs soaking period was given followed by injection of 0.1M NaCl solution at the rate of 0.2 cc/min. The turbidity analysis showed no presence of fine particles in the effluent as the turbidity value was only 1.39 NTU after injecting 4 PVs of low salinity water (Fig. 1a). The UV spectrophotometry results also confirmed the absence of fine particles in the effluent as the measured absorbance was very low (0.07 ABS). This test showed the effectiveness of nanotechnology in mitigating fine migration problems during low salinity waterflooding operations. The DLVO model was also used to further validate the experimental data in the case of nanofluid. For this purpose, the zeta potential of the sand-nanofluid system was measured and is plotted in Figure 3a along with sand-brine zeta potentials. As expected, nanoparticles decreased the absolute value of sand-brine zeta potential, implying higher attractive forces. The DLVO model was then run for the sand-nanofluid system and the results are presented in Figure 3b with the sand-brine cases. It is clear from Figure 3b that the silica nanoparticles successfully altered the surface potential, causing an increase in the attractive forces. Hence, fine particles remained attached to the sandstone surface in the presence of nanofluid, indicating that the CSC for this system was now reduced to a lesser value than 0.1M.

## 4. Conclusions

The estimation of CSC is necessary to avoid formation damage during many oil recovery processes, particularly in low salinity waterflooding and alkaline flooding. From core-flood experiments and effluent analysis, the CSC for sandstone-NaCl brine was in between 0.15M and 0.1M which was also validated through the analysis of surface forces and DLVO modeling. DLVO model results were in close agreement with the experimental results and this tool can be used to estimate the CSC without extensive experimentation. The effect of nanoparticles on fine migration control was also studied and it was found that nanoparticles altered the surface potential and reduced repulsive forces in the SFB system, thereby mitigating the fines migration issue and lowering the CSC. Both the experimental results and DLVO model outcome proved that nanotechnology has the potential to improve the performance of enhanced oil recovery methods by controlling fines production and overcoming productivity and injectivity issues.

## Acknowledgments

The authors would like to thank Nazarbayev University for supporting this research through the NU Faculty Development Competitive Research Grants program (Application of Nanofluids to Control Formation Damage and Improved Oil Recovery Process, Code: 240919FD3928).

## References

[1] Bjørlykke, K. and J. Jahren, *Sandstones and Sandstone Reservoirs*, in *Petroleum Geoscience: From Sedimentary Environments to Rock Physics*. 2010, Springer Berlin Heidelberg: Berlin, Heidelberg. p. 113-140.

[2] Chavan, M., A. Dandekar, S. Patil, and S. Khataniar, *Low-salinity-based enhanced oil recovery literature review and associated screening criteria.* Petroleum Science, 2019; **16(6)**: p. 1344-1360.

[3] Khilar, K.C. and H.S. Fogler, *The existence of a critical salt concentration for particle release.* Journal of colloid and interface science, 1984; **101(1)**: p. 214-224.

[4] Borazjani, S., L. Chequer, T. Russell, and P. Bedrikovetsky, *Injectivity Decline During Waterflooding and PWRI due to Fines Migration*, in *SPE International Conference and Exhibition on Formation Damage Control*. 2018, Society of Petroleum Engineers: Lafayette, Louisiana, USA. p. 11.

[5] You, Z. and P. Bedrikovetsky, *Well Productivity Impairment Due to Fines Migration*, in *SPE International Conference and Exhibition on Formation Damage Control*. 2018, Society of Petroleum Engineers: Lafayette, Louisiana, USA. p. 23.

[6] Khilar, K.C. and H.S. Fogler, *Migrations of fines in porous media.* Vol. **12**. 1998: Springer Science & Business Media.

[7] Mansouri, M., A. Nakhaee, and P. Pourafshary, *Effect of SiO2 nanoparticles on fines stabilization during low salinity water flooding in sandstones.* Journal of Petroleum Science and Engineering, 2019; **174**: p. 637-648.

[8] Assef, Y., D. Arab, and P. Pourafshary, *Application of nanofluid to control fines migration to improve the performance of low salinity water flooding and alkaline flooding.* Journal of Petroleum Science and Engineering, 2014; **124**: p. 331-340.

[9] Díez, R., O.E. Medina, L.J. Giraldo, and F.B. Cortés, *Development of Nanofluids for the Inhibition of Formation Damage Caused by Fines Migration: Effect of the interaction of Quaternary Amine (CTAB) and MgO Nanoparticles.* Nanomaterials, 2020; 10(5): p. 928.

[10] Franco, C.A., R. Zabala, and F.B. Cortés, *Nanotechnology applied to the enhancement of oil and gas productivity and recovery of Colombian fields.* Journal of Petroleum Science and Engineering, 2017; **157**: p. 39-55.

[11] Derjaguin, B. and L. Landau, *The theory of stability of highly charged lyophobic sols and coalescence of highly charged particles in electrolyte solutions.* Acta Physicochimica URSS, 1941; **14**(633-52): p. 58.

[12] Verwey, E.J.W., *Theory of the stability of lyophobic colloids: the interaction of sol particles having an electric double layer*, 1962; V(4): No. 541.3453

[13] Loeb, A.L., J.T.G. Overbeek, P. Wiersema, and C. King, *The electrical double layer around a spherical colloid particle.* Journal of The Electrochemical Society, 1961. 108(12): p. 269C.

[14] Souris, J.S., N.-T. Chen, S.-H. Cheng, C.-T. Chen, and L.-W. Lo, *Chapter 20 - Silica Nanoparticle Platform*, in *Cancer Theranostics*, X. Chen and S. Wong, Editors. 2014, Academic Press: Oxford. p. 363-391.

[15] Muneer, R., M.R. Hashmet, and P. Pourafshary, *Fine Migration Control in Sandstones: Surface Force Analysis and Application of DLVO Theory.* ACS omega, 2020 Dec 2; **5(49)**: 31624-39

IOP Conf. Series: Materials Science and Engineering 1186 (2021) 012002 doi:10.1088/1757-899X/1186/1/012002

# Hydrodynamic response time of magnetorheological fluid in valve mode: an initial study

**Michal Kubík[1, a *] and Karel Šebesta**

[1] Brno University of Technology, Faculty of Mechanical Engineering, Brno, Czechia

[a] michal.kubik@vutbr.cz

**Abstract.** The transient behaviour of magnetorheological (MR) damper is a very important parameter affecting the performance of this technology in modern semi-actively controlled suspension systems. Currently, the transient behaviour of the MR damper is limited by dynamics of the MR fluid (MRF) itself. The significant part of MRF response time is a hydrodynamic response time which is connected with transient rheology and development of velocity profile in the slit gap. In this paper, the method for measuring the hydrodynamic response time of MRF operating in valve mode is presented. The hydrodynamic response time of MRF-132DG achieved value of $\tau_{90} = 0.78$ ms for H = 17.5 kA/m a value of $\tau_{90} = 0.65$ ms for H = 34 kA/m for given geometry of gap. The difference between model and experiment is lower in higher yield stresses of MRF.

## 1. Introduction

Magnetorheological (MR) fluid is a smart material that exhibits a reversible and fast transition from a liquid state to a solid-state (increase of yield stress) under an external magnetic field. This phenomenon is usually called a magnetorheological effect. MR fluid is a two-phase fluid consisting of micron-sized highly magnetizable particles, usually carbonyl iron particles, in a non-magnetizable carrier fluid, such as mineral or silicone oil. When the MR fluid is energized by the magnetic field, the ferromagnetic particles are magnetized and form chain-like structures in the direction of the magnetic field. The fluid then exhibits a significant increase in yield stress. These properties allow the use of MR fluid in electro-mechanical systems such as dampers. Magnetorheological (MR) dampers are currently used in automotive, railway, aviation, etc. The transient (dynamic) behavior of the MR damper is a fundamental property affecting the performance of this technology in modern semi-actively controlled suspension systems. Two sources of the time delay between the control signal and damping force are as follow: (i) dynamics of MR damper hardware and (ii) the magnetorheological fluid dynamics.

## 2. Dynamics of MR damper hardware

The response time of MR damper hardware was listed in several publications [1,2] and the main sources of the delay were identified as (i) eddy currents in coil core, (ii) inductance of MR damper coil, or (iii) compressibility of the hydraulic system of MR damper. When the magnetic field changes rapidly in the magnetic circuit of the damper, eddy currents are generated, which produce a magnetic field in the opposite direction than the initializing magnetic field [3]. Eddy currents can be eliminated by either a material (SMC, ferrite, etc.) or shape approach. The shaped approach is based on the use of a magnetic circuit with thin sheets or with suitably located grooves. This issue is currently being under development. The relatively slow increase and decrease of electric current on control voltage signal due to the inductance of the coil can be solved by the current controller with the over-voltage method. The compressibility of the magnetorheological fluid also affects the transient behaviour of the MR damper [4]. The issue of MR damper hardware dynamics is relatively intensively described. Currently published MR dampers achieve response time (1.2 ms) comparable with a response time of MR fluid itself [5].

## 3. Dynamics of MR fluid itself

The transient (dynamic) behavior of MR fluid was described just in a few papers. The best-known paper in connection with the MR fluid response time is the paper of Goncalves et al. [6]. This team published a measurement of MR fluid response time on a slit-flow rheometer. The author introduces the fluid dwell time. This is the time that ferromagnetic particles of MR fluid spend in the presence of a magnetic field. The author considered that if the dwell time is short enough, the ferromagnetic particles in the MR fluid will not be sufficient to create chains and the yield stress of the MR fluid will not be created. This can be connected with MR fluid particle chain dynamics. The yield stress decrease was observed in Goncalves experiments. However, the papers of Sherman [7] or Goldasz et. al. [8] show that this pressure drop decrease is due to transient rheology connected with the development of velocity profile. This statement is based on CFD simulations. It is assumed that the yield stress of MR fluid flow will decrease/increase much faster than energy dissipation (pressure drop) due to flow transient. This phenomenon is often referred to as the hydrodynamic fluid response time. Sherman [7] published a model for the hydrodynamic response of MR fluids. This model is based on an analytical solution of the start-up flow of Bingham plastic fluid between parallel plates [9]. The Sherman [7] construct fit of data Darpa et al. [9] (dimensionless response time $T_r$) as follows:

$$T_r = 0.235/(1+0.2B_i)$$

where $B_i$ is Bingham number. For a Newtonian fluid, the dimensionless response time can be calculated as follows:

$$T_r = \tau_r \eta / \rho h^2$$

where $\tau_r$ response time, $\eta$ is plastic viscosity, $\rho$ is the density of the fluid, $h$ is gap size. It is important to note that this is the time $\tau_r$ to reach 90% of the final steady-state velocity.

The main aim of this paper is to present a method for measuring the hydrodynamic response time of MR fluid operating in valve mode and preliminary results. The measured data will be compared with the model published by Sherman [7].

## 4. Materials and methods

### 4.1. Measurement method idea

This method is based on the measurement of pressure drop of the MR valve caused by an external magnetic field acting on the MR fluid $\Delta p_\tau$ depending on dwell time $t_{dwell}$. This method significantly extend the paper of Goncalves et al. [6]. The challenge is to separate the effect of viscosity on pressure drop from measured data and it change with temperature. Therefore, the measurement procedure consists of two blocks. The first measurement block is with zero magnetic fields (H = 0 kA/m) and Newtonian behavior of the magnetorheological fluid is considered in area I to III, see Figure. 1. The pressure drop $\Delta p_{off}$ (H = 0 kA/m) for the first section (I) can be described as follows:

$$p_I - p_{III} = \Delta p_{off} = \Delta p_\eta + \Delta p_\chi$$

where $\Delta p_\eta$ is the viscous pressure drop and $\Delta p_\chi$ is the pressure drop of entrance and exit losses to the gap. In the second section II of measurement, the test MR valve (gap) is divided into two areas: (I-II) without a magnetic field and (II-III) with a magnetic field, see Figure. 1. It is assumed that the velocity profile on interface (II) is fully developed (Newtonian).

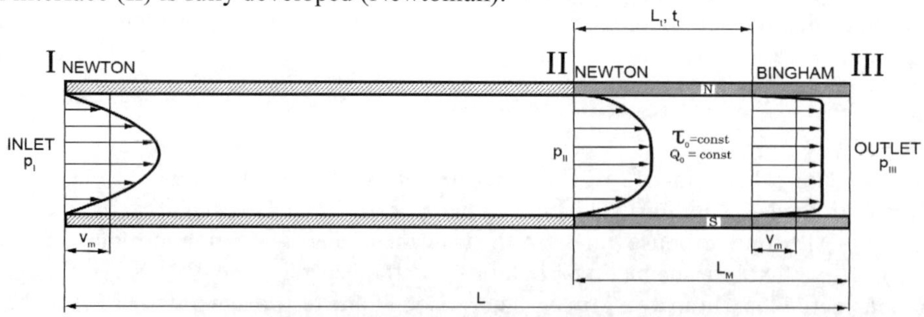


**Figure. 1** MR fluid flow in slit gap

The second measurement of the pressure loss $\Delta p_{on}$ is already in the presence of a magnetic field ($H \neq 0$ kA/m) at a given length ($L_M$). The pressure drop $\Delta p_{on}$ is then:

$$p_I - p_{III} = \Delta p_{on} = \Delta p_\eta + \Delta p_\chi + \Delta p_\tau = \Delta p_{off} + \Delta p_\tau$$

where $\Delta p_\tau$ is pressure drop due to yield stress of MR fluid on the length $L_M$. From the measured pressure drops $\Delta p_{off}$ and $\Delta p_{on}$, the pressure drop caused by the MR fluid yield stress $\Delta p_\tau$ is determined as follows:

$$\Delta p_\tau = \Delta p_{on} - \Delta p_{off}$$

These two sections must be measured as quickly as possible to keep the temperature difference as small as possible. However, there is always some change in temperature, which is reflected in the change in viscosity and is compensated in data post processing. Next, the dwell time $t_{dwell}$ was defined. This is the time that the MR fluid spends in the presence of a magnetic field. The dwell time $t_{dwell}$ was calculated by the mean flow velocity in the gap $v_m$ and the length $L_M$ as follows:

$$t_{dwell} = v_m / L_M$$

It is necessary to note that the length $L_M$ is not connected with magnetic pole geometry but has to be calculated by magnetic model. In our case $L_M$ achieved a value of 8.8 mm. It is assumed that the MR fluid needs a certain time $t_t$ or length $L_t$ to transform the Newtonian velocity profile to a velocity profile with a plug (Bingham plastic) to exhibits dissipation energy due to MR fluid yield stress. This time $t_t$ or length $L_t$ is probably the result of MR fluid deceleration/ acceleration in the gap. In the case of $L_t >> L_M$ and at given mean velocity in the gap $v_m$, the velocity profile was not developed from Newtonian to Bingham, and dissipation energy due to MR yield stress is not observed $\Delta p_\tau = 0$ although the yield stress is not zero ($H \neq 0$ kA/m). Otherwise ($L_t << L_M$) the velocity profile was developed very quickly and there is no significant effect of MR fluid deceleration/acceleration.

*4.2. Experimental device*

The functional layout of the apparatus is shown in Figure. 2. The experimental device (slit-flow rheometer) is energized by the Inova hydraulic dynamometer which imposes the motion of the floating piston ($d_p = 32$ mm). The motion of the floating piston forces the test MR fluid through the slit gap of the rheometer (MR valve). The slit gap geometry (gap size $t_g = 0.6$ mm, width $w_g = 32$ mm and length $L = 50$ mm) is made of two grind blocks (position 4 a 5 in Figure. 3). Each block consists of a part made of non-magnetic stainless steel (position 5.1) and a part made of pure iron (5.2). Therefore, the magnetic flux can flow just through part 5.2.

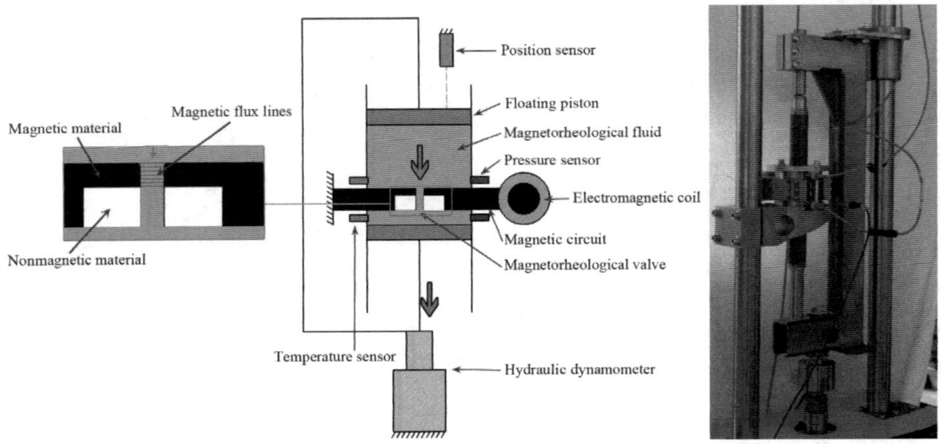


**Figure. 2** Functional layout of the rheometer, rheometer during tests on a hydraulic dynamometer

The magnetic circuit is composed of the electromagnetic coil (1), pole pieces (2), and non-magnetic washer (3). The magnetic pole pieces are manufactured from pure iron. The electromagnetic coil is wound with 160 turns of copper wire with a diameter of 0.5 mm. The important dimensions are shown in Figure. 3.

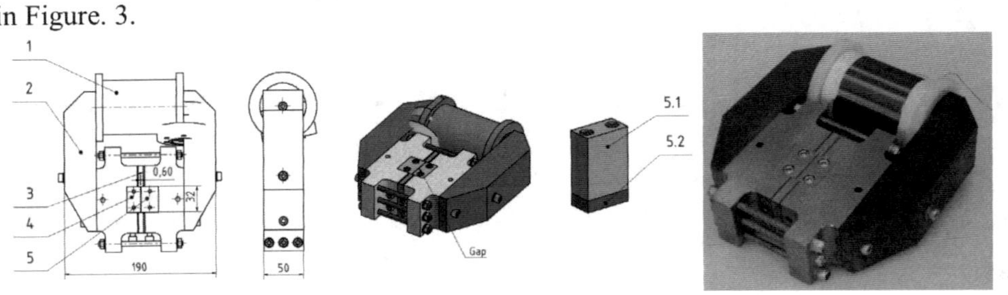


**Figure. 3** Important dimension of magnetic circuit (left, middle), manufactured magnetic circuit (right)

*4.3. Measurement methodology*

The pressure drop across the slit gap is acquired with pressure sensors HBM P8AP. The MR fluid temperature was acquired by resistance temperature sensor PT100 which was located near the entrance to the slit gap. The motion of the floating piston was measured using the LVDT transducer/position sensor accommodated in a hydraulic dynamometer. These signals were recorded and conditioned with a sampling frequency of 1 kHz by analyzer Dewe-800. The measurement procedure was as follow: (i) 8 sinusoidal cycles at given frequency *f* were measured; (ii) pause of 10 sec to switch on the power supply current *I*; (iii) 8 sinusoidal cycles at given frequency *f* and given electric current *I* were measured; (iv) demagnetization of the magnetic circuit using a sinusoidal cycle with decreasing amplitude. The rise and fall of the frequency were always with a transition time of 2 s. Measurements were performed for different frequencies at a stroke amplitude of 48 mm. This corresponds to a maximum piston velocity of 0.48 m/s and a mean flow velocity in the slit gap of 22 m/s. The piston velocity $v_m$ was determined by derivation of piston position. The data of pressure drop from the center of stroke (± 2 mm) were selected for the next evaluation. The measured data which achieved at least 99 % of maximum piston velocity was taken also for the next evaluation to eliminate transition to given frequency *f*. Selected points of pressure drop $\Delta p_{off}$ and pressure drop $\Delta p_{on}$ were averaged and presented in the chapter results. The results were normalized $\psi$ due to the quantitative comparison. The normalized data $\psi$ were then fitted by a least-squares method in Matlab software by the following function:

$$\psi(t_{dwell}) = \left[1 - c_1 . e^{c_2 .(t_{dwell} - c_4)}\right] . tanh(c_3 . (t_{dwell} - c_4)$$

IOP Conf. Series: Materials Science and Engineering 1186 (2021) 012002 doi:10.1088/1757-899X/1186/1/012002

where $c_1$, $c_2$, $c_3$, and $c_4$ are constants that were obtained by the fitting. From this equation, the primary ($\tau_{63}$) and secondary ($\tau_{90}$) hydrodynamic responses of the MR fluid were determined. MR fluid MRF-132DG from Lord Corporation was used for testing. The volume concentration of ferromagnetic particles in the tested sample was experimentally determined to be 32.6 % and the density to be 3106 kg/m$^3$.

## 5. Results and discussion

The graphs below (Figure. 4) show the dependences of the pressure drop $\Delta p_{on}$ and $\Delta p_{off}$ on the mean flow velocity in the slit gap $v_m$ for an electric current of 0.4 A and 0.6 A applied to the electromagnetic coil.

It can be seen that with increasing gap velocity $v_m$, the differences in pressure drop decrease. This also corresponds to the data published by [10].

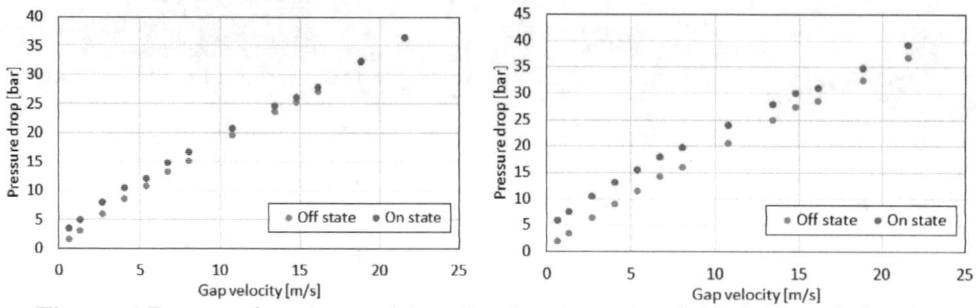


**Figure. 4** Pressure drop $\Delta p_{on}$ and $\Delta p_{off}$ for electric current 0.4A (left), and electric current 0.6 A (right).

The pressure drop $\Delta p_\tau$ was then calculated as a function of dwell time $t_{dwell}$, and the data were normalized, see Figure. 5. The measured data were fitting functions presented above.

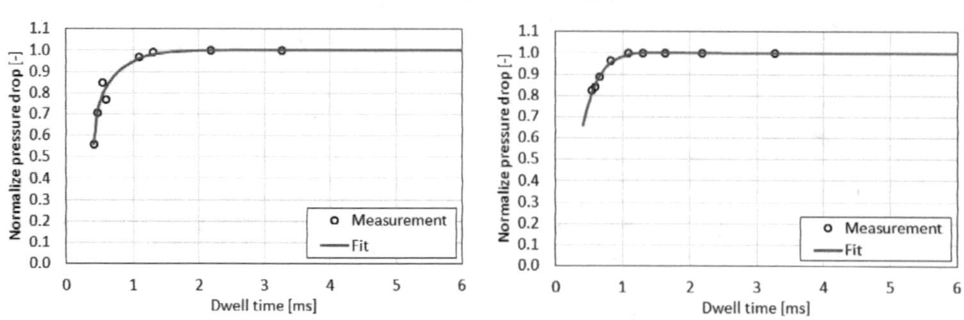


**Figure. 5** Normalize pressure drop $\Delta p_\tau$ for electric current 0.4 A (left), and for electric current 0.6 A (right)

The table below (Table 1) shows the fitting coefficients $c_1$ to $c_4$, primary and secondary response time from the experiments for the two magnetic fields. The magnetic flux intensity *H* was estimated based on data from the MR fluid manufacturer and measured yield stress. The results of the experiment were compared with the model.

**Table 1.** Result of the experiments

| I [A] | $c_1$ [-] | $c_2$ [-] | $c_3$ [-] | $c_4$ [-] | $\tau_{63}$ [ms] | $\tau_{90}$ [ms] | $\tau_{r\,m}$[ms] | Y. stress [kPa] | H [kA/m] |
|---|---|---|---|---|---|---|---|---|---|
| 0.4 | 0.38 | -2.98 | 0.332 | 14.75 | 0.43 | 0.78 | 1.15 | 5.9 | 17.5 |
| 0.6 | -0.15 | -2.31 | 0.09 | 2.28 | 0.4 | 0.65 | 0.9 | 12.8 | 34 |

For the MR fluid yield stress of 5.9 kPa (17.5 kA/m), the secondary response time was $\tau_{90} = 0.78$ ms which is lower than predicted by model ($\tau_{r\ model} = 1.15$ ms). In 5this case, the difference between model ad experiemnt is 34 %. At a higher value of yield stress (12.8 kPa), the difference between model and expeiemnt is slightly lower (28 %). Nevertheless, it can be stated that the measured data corresponding to the model. The differences can be explained by slightly different temperature for each measured point (each dwell time) and simplification of the model. In the following research, it is necessary to focus on determining the effect of the magnetic field, density, or temperature on measured hydrodynamic response time.

## 6. Conclusion

In this paper, the measurement method of the hydrodynamic response time of magnetorheological fluid was presented. The measured data of hydrodynamic response time of MRF-132DG for two levels of the magnetic field was compared with the model. The difference between model and experiment is lower in higher yield stresses of MRF. The measured hydrodynamic response time for yield stress of 12.8 kPa (H = 34 kA/m) was 28 % lower than predicted by the model and achieved a value of 0.65 ms. It can be stated that the model published Sherman correlate with our initial experiment for given geometry and MR fluid rheology. The model is simplified and therefore a difference of about 30% is acceptable.

## Acknowledgments

This paper and experiments have been made possible thanks to the Czech Science Foundation (GAČR) No. 20-23261Y.

## References

[1] Koo, J.H.; Goncalves, F.D.; Ahmadian, M. A comprehensive analysis of the response time of MR dampers. *Smart Mater. Struct.* **2006**, *15*, 351–358.

[2] Kubík, M.; Macháček, O.; Strecker, Z.; Roupec, J.; Mazůrek, I. Design and testing of magnetorheological valve with fast force response time and great dynamic force range. *Smart Mater. Struct.* **2017**, *26*, 1–9.

[3] Strecker, Z.; Mazůrek, I.; Roupec, J.; Klapka, M. Influence of MR damper response time on semiactive suspension control efficiency. *Meccanica* **2015**, *50*, 1949–1959.

[4] Kubík, M.; Goldasz, J. Multiphysics Model of an MR Damper including Magnetic Hysteresis. *Shock Vib.* **2019**, *2019*.

[5] Strecker, Z.; Roupec, J.; Mazurek, I.; Machacek, O.; Kubik, M.; Klapka, M. Design of magnetorheological damper with short time response. *J. Intell. Mater. Syst. Struct.* **2015**, *26*, 1951–1958.

[6] Goncalves, F.D.; Ahmadian, M.; Carlson, J.D. Investigating the magnetorheological effect at high flow velocities. *Smart Mater. Struct.* **2005**, *15*, 75–85.

[7] Sherman, S.G. MAGNETORHEOLOGICAL FLUID DYNAMICS FOR HIGH SPEED ENERGY ABSORBERS, 2017.

[8] Gołdasz, J.; Sapiński, B. *Insight into magnetorheological shock absorbers*; 2015; ISBN 978-3-319-13232-7.

[9] Dapri, I.; Scarpi, G. Start-up Flow of a Bingham Fluid in a Pipe. *Meccanica* **2005**, *40*, 49–63.

[10] Mao, M.; Hu, W.; Choi, Y.T.; Wereley, N.; Browne, A.L.; Ulicny, J. Experimental validation of a magnetorheological energy absorber design analysis. *J. Intell. Mater. Syst. Struct.* **2014**, *25*, 352–363.

IOP Conf. Series: Materials Science and Engineering 1186 (2021) 012003 doi:10.1088/1757-899X/1186/1/012003

# Enhanced absorption and EQE of Si MSM photodiodes with integrated periodic arrays of holes

**Jun Gou*, Zheyuan Xie, He Yu, Xianchao Liu, Jun Wang**

State Key Laboratory of Electronic Thin Films and Integrated Devices, University of Electronic Science and Technology of China, Chengdu 610054, China

*goujun@uestc.edu.cn

**Abstract**. In order to enhance the absorption of thin Si film, cylindrical-shaped holes arranged in square and hexagonal lattices with different designs of hole diameter, depth and period are integrated in the surface of Si films with a thickness of 1.5 μm and 2.5 μm on either bulk Si or SOI substrate. At the wavelengths between 800 ~ 950 nm, light bending and absorption enhancement of Si film with integrated hole array is analyzed by FDTD simulation. Hole array with a period similar to the wavelength and a diameter/period ratio around 0.7 enables a higher absorption of Si film. Si film with cylindrical-shaped holes in hexagonal lattice with a Si thickness of 1.5 μm and a hole depth of 250 nm provides an absorption of 43.16 % at 850 nm wavelength on a $SiO_2$ substrate. Nanostructured Si MSM photodiodes are fabricated by integrating periodic arrays of holes in 1.5 μm thin absorption regions between metal fingers. The EQE is enhanced to 61 % at 850 nm, while the photodiodes without holes show only 23 %.

## 1. Introduction

CMOS-process-compatible silicon (Si) photodiodes could lead to a cheaper and better integrated fibre-optic communication system [1]. However, Si photodiodes are currently used predominantly in the visible wavelength regime. At near-infrared (IR) wavelengths such as 840 ~ 860 nm used for short-reach (< 300 m) data communication, traditional Si PIN photodiodes with a weak absorption coefficient of Si can hardly achieve a high external quantum efficiency (EQE, >50%) and a high data rate (20 Gb/s) simultaneously because the former needs a Si thickness of >13 μm and the latter restricts the thickness to < 2.5 μm [2]. One effective approagh to enhance the absorption of a thin Si film is to integrate micro- and nanostructures in the Si film with the purpose of tailoring its optical properties and trapping the photons [3]. Such schemes include the formation of nanowires [4], nanodomes [5], nanodiscs [6], nanopencils [7], and so on. In recent studies, an all-Si PIN photodiode fabricated on a silicon-on-insulator (SOI) wafer with periodic funnel-shaped holes is presented for efficient light trapping, resulting in an enhanced EQE of 52% at 850 nm with a 2-μm-thick *i*-Si layer and an ultra-fast impulse response (full-width at half-maximum) of 30 ps [8]. It reveals that a beam of vertically incident light is guided laterally and confined in the nanostructured Si film until it gets absorbed [9]. Compared to typical PIN photodiodes, MSM photodiodes with a simple structure of interdigitated metal fingers on Si can provide a high response speed due to its low capacitance which leads to a low RC time. However, Si MSM photodiodes also have a low responsivity at near-IR wavelengths between 800 ~ 950 nm.

In this work, in order to increase the EQE of Si MSM photodiodes while maintaining a high-speed operation, periodic arrays of photon-trapping holes are integrated in thin absorption regions between metal fingers. Cylindrical-shaped holes arranged in square and hexagonal lattices with different designs of hole diameter, depth and period are designed in the surface of Si films with a thickness of 1.5 μm and 2.5 μm on either bulk Si or $SiO_2$ substrate. FDTD simulation is used to investigate the absorption of nanostructured Si film at the wavelengths between 800 ~ 950 nm. The absorption of a 1.5 μm thin Si film on a $SiO_2$ substrate is enhanced to 43.16 % at 850 nm by integrating periodic cylindrical-shaped holes with a depth of 250 nm. The EQE of Si MSM photodiodes is increased to 61 % compared to photodiodes in the absence of photon-trapping hole structures which shows an EQE of 23 %.

## 2. Design and fabrication

The Si photodiode structure is epitaxially grown on either bulk Si wafer with a thickness of about 500 μm or SOI wafer with a thin device layer over a thick $SiO_2$ layer. Here, Si films with a thickness ($t$) of 1.5 μm and 2.5 μm are designed on a bulk Si substrate as shown in Fig. 1(a) and a 3-μm-thick $SiO_2$ film over a bulk Si substrate as shown in Fig. 1(b). Micro- and nanoscale periodic cylindrical-shaped holes with different designs of hole diameter ($d$), depth ($h$) and period ($p$) are integrated in the surface of Si films. Hole arrays are arranged in square lattice as shown in Fig. 1(c) and hexagonal lattice as shown in Fig. 1(d), respectively. FDTD simulation is used to model the nanostructured Si films with different designs of hole array and substrate, as shown in Fig. 1(e). Light reflection $R(\lambda)$ from the top surface and transmission $T(\lambda)$ into the substrate are simulated, and thus the absorption can be calculated by $A(\lambda) = 1 - R(\lambda) - T(\lambda)$ at the wavelengths between 800 ~ 950 nm. Light propagation and electric field distribution in the Si film with hole array are also analyzed.

Si MSM photodiodes were fabricated on a SOI substrate that has 1.5 μm device layer (p-Si) and 3 μm BOX layer. 100 nm thick interdigitated aluminum (Al) fingers were fabricated on Si by sputtering and lift-off processes. Afterwards, periodic holes with different diameters and periods were patterned and etched by electron beam lithography and reactive ion etching to form 0.25 μm deep cylindrical holes in Si. A device without holes was also fabricated in the same process as a comparison. Five different designs of diameter ($d$), period ($p$), and arrangement of holes are shown in Table 1. The width ($w$) and spacing ($s$) of Al fingers are also shown in the table. Fig. 1 (e) shows the scanning electron microscopy (SEM) image of a fabricated device with holes with a diameter/period ($d/p$) of 700/1000 nm.

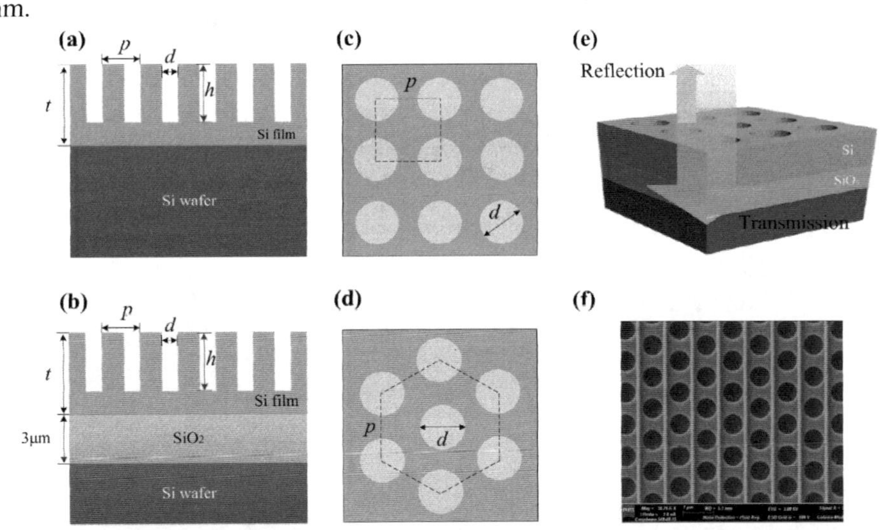


**Figure 1.** Design of nanostructured Si film and fabricated MSM photodiode with integrated hole array. (a and b) Si film with hole array on a bulk Si wafer (a) and a SOI wafer with a 3-μm-thick $SiO_2$ film (b), showing cylindrical-shaped holes integrated in the surface of Si film. (c and d) Holes arranged in square lattice (c) and hexagonal lattice (d). (e) FDTD model of the holes integrated Si film on a SOI

substrate illuminated by a vertical incident light. (f) SEM image of a fabricated device with holes with a diameter/period (*d/p*) of 700/1000 nm.

**Table 1.** Si MSM photodiodes with different designs of holes and fingers

| | *d/p* (nm) | arrangement | *w* (nm) | *s* (nm) |
|---|---|---|---|---|
| 1# | - | - | 150 | 740 |
| 2# | 700/1000 | square lattice | 150 | 740 |
| 3# | 630/900 | hexagonal lattice | 130 | 670 |
| 4# | 700/1000 | hexagonal lattice | 150 | 740 |
| 5# | 1400/2000 | hexagonal lattice | 300 | 1440 |

## 3. Results and discussion

### *3.1. Optimized absorption of Si film with integrated photon-trapping holes*

In order to figure out the relationship between optical absorption and photodiode structure with integrated holes, we have done a series of comparative simulation based on FDTD. Fig. 2(a) shows the absorption of a 1.5-μm-thick flat Si film and Si films integrated with hole arrays on bulk Si and $SiO_2$ substrates at the wavelengths between 800 ~ 950 nm. Hole arrays are integrated in the surface of Si films in square and hexagonal lattices with a diameter/period (*d/p*) of 700/1000 nm and a depth of 250 nm. Fig. 2(a) reveals that the flat Si film has a very low absorption (< 10%) on either bulk Si or $SiO_2$ substrate at 800 ~ 950 nm wavelengths. The absorption of the flat Si film on a $SiO_2$ substrate is a little higher than that of the Si film on a bulk Si substrate due to the back reflection of $SiO_2$ layer. The absorption of Si film is enhanced by integrating periodic holes in the surface. For Si film on a bulk Si substrate, the absorption is improved to 10% ~ 20% at the wavelengths between 800 ~ 950 nm and the absorption of Si films with different hole arrangements has little difference. For Si film on a $SiO_2$ substrate, the absorption is significantly enhanced by integrating hole array. Taking the data communication wavelength of 850 nm for an example, the absorption of the flat Si film with a thickness of 1.5 μm on bulk Si and $SiO_2$ substrates are 3.66% and 6.9%, respectively. Si film with hole array on a bulk Si substrate has an absorption of 11.5%, whereas the absorption of Si film with the same thickness on a $SiO_2$ substrate is enhanced to 35.43% for square hole lattice and 43.16% for hexagonal hole lattice. This represents over 1100% improvement compared to Si film on a bulk Si substrate without integrated nanoholes. Generally, Si film with holes in hexagonal lattice has a higher absorption than its counterpart with holes in square lattice due to a higher porosity and lower surface reflection with the same diameter and period of holes. At the wavelengths between 800 ~ 950 nm, Si film with hole array in hexagonal lattice provides a high absorption of 30.45%~47.53%.

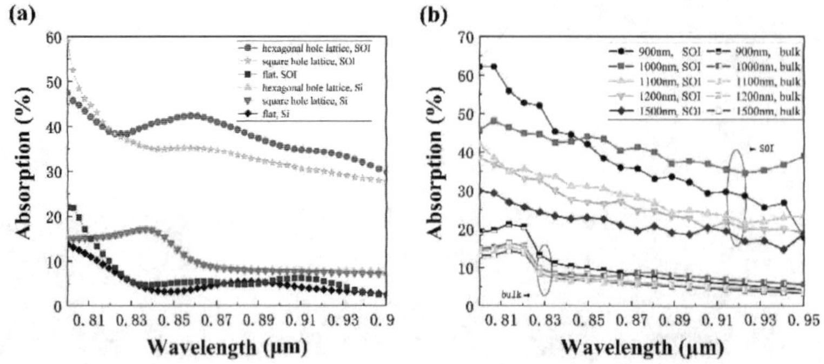


**Figure 2.** Absorption of 1.5-μm-thick holes integrated Si films with different designs of hole arrangement, substrate and period of holes. (a) Absorption of a 1.5-μm-thick flat Si film and Si films with hole arrays on bulk Si and $SiO_2$ substrates at the wavelengths between 800 ~ 950 nm. Hole arrays are integrated in the surface of Si films in square and hexagonal lattices with a diameter/period (*d/p*)

of 700/1000 nm and a depth of 250 nm. (b) absorption of 1.5-μm-thick Si films with hole arrays in hexagonal lattice with different periods of holes on bulk Si and $SiO_2$ substrates at the wavelengths between 800 ~ 950 nm. Hole arrays have a hole diameter of 700 nm and a depth of 250 nm.

Fig. 2(b) shows the absorption of 1.5-μm-thick Si films with integrated hole arrays in hexagonal lattice with different periods of holes on bulk Si and $SiO_2$ substrates at the wavelengths between 800 ~ 950 nm. Hole arrays have a hole diameter of 700 nm and a depth of 250 nm. It shows that the absorption of Si films with hole arrays on a $SiO_2$ substrate outperform their counterparts with the same hole structures on a bulk Si substrate. When the period of holes increases from 900 nm to 1500 nm, higher absorption is achieved for the Si film with smaller periods of holes ($p$ = 900 nm and $p$ = 1000 nm, $d/p$ = 0.78 and $d/p$ = 0.7) on either bulk Si or $SiO_2$ substrate. The absorption decreases with increasing period of holes. Holes integrated Si film with a period of 1000 nm has a lower absorption at the wavelengths of 800 ~ 850 nm and higher absorption at the wavelengths of 850 ~ 900 nm compared to holes integrated Si film with a period of 900 nm.

Fig. 3(a) and Fig. 3(b) shows the absorption of 1.5-μm-thick and 2.5-μm-thick Si films with hole arrays in hexagonal lattice with different hole depths on a $SiO_2$ substrate at the wavelengths between 800 ~ 950 nm, respectively. Hole arrays are integrated with a diameter/period ($d/p$) of 700/1000 nm. Absorption of the flat Si film without holes with the same thickness is also presented. Fig. 3(a) reveals that when the holes span the whole Si film ($h$ = 1.5 μm), although the absorption is higher than that of the flat Si film, the nanostructured Si film has the lowest absorption compared to Si films with hole arrays of other depths ($h$ < 1.5 μm). The highest absorption is obtained with a hole depth of 250 nm. Fig. 3(b) shows that for the 2.5-μm-thick Si film, the absorption of holes integrated Si film with a hole depth of 2.5 μm is also the lowest compared to other hole depths, and 250 nm is still a hole depth that can provide a high absorption especially for the longer wavelengths between 840 ~ 950 nm. A high absorption can also be obtained by the Si film with hole array with a hole depth of 2 μm which provides a higher absorption at the wavelengths of 800 ~ 840 nm. It can be concluded that it is advantageous for light absorption to keep a layer of flat Si film without holes with a certain thickness between $SiO_2$ substrate and the hole array, which is helpful to guide the light in laterally propagating modes in the Si film.

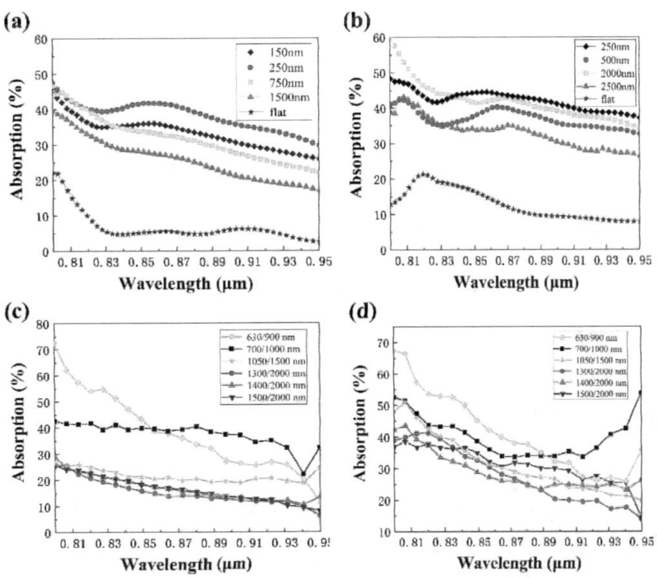


**Figure 3.** (a and b) Absorption of 1.5-μm-thick (a) and 2.5-μm-thick (b) Si films with hole arrays in hexagonal lattice with different hole depths on a $SiO_2$ substrate at the wavelengths between 800 ~ 950 nm. Hole arrays are integrated with a diameter/period ($d/p$) of 700/1000 nm. (c and d) Absorption of 1.5-μm-thick Si films with a hole depth of 250 nm (c) and 2.5-μm-thick Si films with a hole depth of 2

μm (d) with six different designs of diameter (*d*) and period (*p*) at the wavelengths between 800 ~ 950 nm

Hole arrays with similar diameter/period (*d/p*) around 0.7 and different designs of hole diameter and period are integrated in Si films on a $SiO_2$ substrate. Absorption of 1.5-μm-thick Si films with a hole depth of 250 nm and 2.5-μm-thick Si films with a hole depth of 2 μm at the wavelengths between 800 ~ 950 nm are shown in Fig. 3(c) and Fig. 3(d), respectively. The holes have six different designs of diameter (*d*) and period (*p*): (I) $d$ = 630 nm, $p$ = 900 nm ($d/p$ = 0.7), (II) $d$ = 700nm, $p$ = 1000 nm ($d/p$ = 0.7), (III) $d$ = 1050 nm, $p$ = 1500 nm ($d/p$ = 0.7), (IV) $d$ = 1300 nm, $p$ = 2000 nm ($d/p$ = 0.65), (V) $d$ = 1400 nm, $p$ = 2000 nm ($d/p$ = 0.7), (VI) $d$ = 1500 nm, $p$ = 2000 nm ($d/p$ = 0.75). It shows that smaller periods of holes ($p$ = 900 nm and $p$ = 1000 nm) lead to a higher absorption of Si film regardless of Si film thickness and hole depth. The absorption decreases as the period of holes increases, which is consistent with the results shown in Fig. 2 (b). Si film with hole array with a period of 900 nm has a higher absorption at the lower wavelengths while Si film with hole array with a period of 1000 nm has a higher absorption at the higher wavelengths. It can be concluded that a higher absorption of the holes integrated Si film can be obtained with a diameter/period (d/p) around 0.7 and a smaller period of holes similar to the wavelength.

### *3.2. Laterally propagating modes of light in Si film with holes*

In order to verify the light bending effect of hole array, electric field distribution in nanostructured Si films with $d/p$ = 700/1000 nm and $d/p$ = 1500/2000 nm are simulated and shown in Fig. 4 at the wavelength of 850 nm. The holes arranged in square lattice with a depth of 1 μm are integrated in the surface of Si films with a thickness of 1.5 μm on a $SiO_2$ substrate. Light is vertically illuminated at one point on the Si surface between two holes with a coordinate value of 0 at the time of $t$ = 0 fs. Electric field distribution in nanostructured Si films at $t$ = 5 fs, $t$ = 7 fs, $t$ = 9 fs, $t$ = 11 fs and $t$ = 20 fs shows light propagation over time in the Si films with different designs of hole array. It can be seen that the vertically incident light undergoes a bending and propagates laterally for each nanostructured Si film due to the diffraction of hole array. It shows a strong electric field distribution in the holes, and the effective optical path of light in Si film is also increased with an enhanced absorption. In the Si film with hole array with $d/p$ = 700/1000 nm, a strong electric field distribution can be observed in the holes 2 μm away from the incident point at $t$ = 5 fs. For the Si film with hole array with $d/p$ = 1500/2000 nm, more than 11 fs is needed to form a strong electric field in the holes at the same distance of 2 μm. Light has a higher laterally propagating speed in the nanostructured Si film with a smaller period of holes.

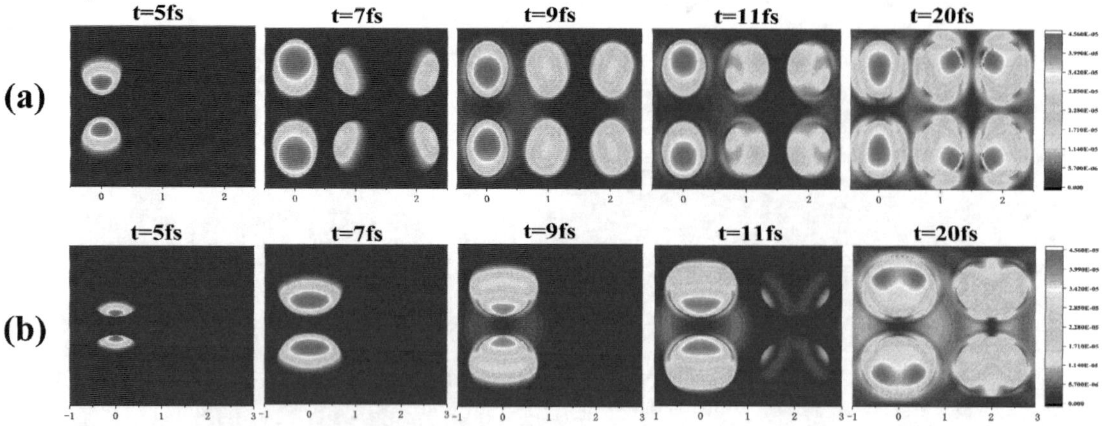


**Figure 4.** Electric field distribution over time in nanostructured Si films with $d/p$ = 700/1000 nm (a) and $d/p$ = 1500/2000 nm (b) at the wavelength of 850 nm. The holes arranged in square lattice with a depth of 1 μm are integrated in the surface of Si films with a thickness of 1.5 μm on a $SiO_2$ substrate.

Light is vertically illuminated at one point on the Si surface between two holes with a coordinate value of 0 at the time of $t = 0$ fs.

### *3.3. Enhanced EQE of Si MSM photodiodes with photon-trapping holes*

The electrical and optical characterization of fabricated Si MSM photodiodes as shown in Table 1 were measured at the wavelength of 850 nm. Fig. 5(a) shows the IV curves of the photodiodes without holes and with holes with $d/p$ = 700/1000 nm under dark and under illumination. The dark current of the photodiodes is lower than 1 μA. An increase in photocurrent can be observed in photodiodes with holes. Fig. 5(b) shows the simulated absorption and measured EQE of the five fabricated photodiodes. The photodiode without holes exhibits an EQE of 23%. By integrating photon-trapping holes, the EQE is significantly increased. The EQE of photodiodes with holes in hexagonal lattice outperform their counterparts with holes in square lattice. When the period of holes in hexagonal lattice decreases from 2000 nm to 900 nm, higher EQE is achieved for the photodiodes with smaller periods of holes ($p$ = 900 nm and $p$ = 1000 nm). This is consistent with the simulation results. The maximum EQE is 61% measured from a photodiode with holes with $d/p$ = 630/900 nm.

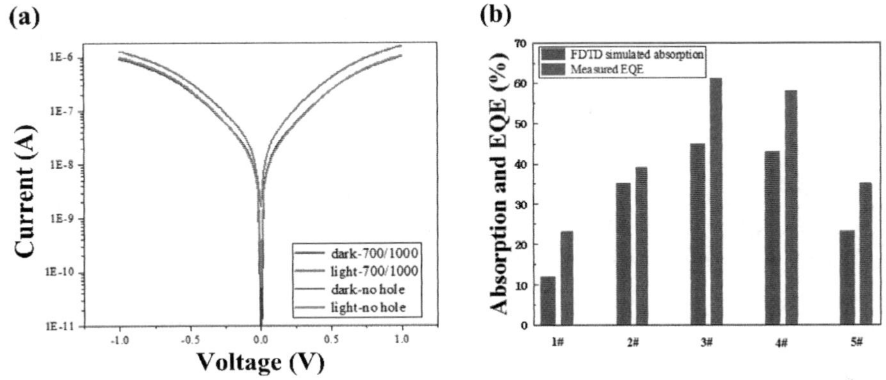


**Figure 5.** Electrical and optical characterization of fabricated Si MSM photodiodes. (a) IV curves of the photodiodes without holes and with holes with $d/p$ = 700/1000 nm under dark and under illumination. (b) Simulated absorption and measured EQE of the five fabricated photodiodes with different designs of holes and fingers.

## 4. Conclusions

In this paper, we have presented a theoretical and experimental study on enhanced absorption and EQE of Si MSM photodiodes with periodic cylindrical-shaped holes at the wavelengths between 800 ~ 950 nm. Hole arrays, designed with different diameter, depth and period, are arranged in square and hexagonal lattices in Si films with a thickness of 1.5 μm and 2.5 μm on bulk Si and $SiO_2$ substrates. Absorption and electric field distribution in nanostructured Si films are simulated to analyze the light bending effect of hole arrays. Si film with hole array with a diameter/period ($d/p$) around 0.7 and a smaller period comparable to the wavelength provides a higher absorption due to more bending of light. For the holes integrated Si film on a $SiO_2$ substrate, light is more trapped and absorbed due to multiple reflections and diffractions, resulting in an enhanced absorption of 30.45% ~ 47.53% at 800 ~ 950 nm wavelengths with a Si thickness of 1.5 μm and a hole depth of 250 nm. EQE of Si MSM photodiodes with photon-trapping holes is enhanced to 61% at 850 nm which represents 262% improvement over the photodiodes without holes.

**Acknowledgements**
This work was supported by National Natural Science Foundation of China (Grant No. 61875031).

**References**

[1] J. A. Tatum, D. Gazula, L. A. Graham, J. K. Guenter, R. H. Johnson, J. King, C. Kocot, G. D. Landry, I. Lyubomirsky, A. N. MacInnes, E. M. Shaw, K. Balemarthy, R. Shubochkin, D. Vaidya, M. Yan, and F. Tang, "VCSEL-based interconnects for current and future data centers," J. Lightwave Technol. 33(4), 727–732 (2015).

[2] M. A. Green, and M. J. Keevers, "Optical properties of intrinsic silicon at 300 K," Prog. Photovoltaics 3, 189−192 (1995).

[3] Z. M. Zhang, Nano/microscale heat transfer. New York: McGraw-Hill, 2007.

[4] E. Garnett, and P. Yang, "Light trapping in silicon nanowire solarcells," Nano Lett. 10, 1082–1087 (2010).

[5] J. Zhu, C.-M. Hsu, Z. Yu, S. Fan, and Y. Cui, "Nanodome solarcells with efficient light management and self-cleaning," Nano Lett.10, 1979–1984 (2010).

[6] C.Rockstuhl, and F. Lederer, "Photon management by metallic nanodiscs in thin film solar cells," Appl. Phys. Lett. 94(21), 213102 (2009).

[7] H. Lin, H. Y. Cheung, F. Xiu, F. Wang, S. P. Yip, N. Han, T. F. Hung, J. Zhou, J. C. Ho, and C. Y. Wong, "Developing controllable anisotropic wet etching to achieve silicon nanorods, nanopencils and nanocones for efficient photon trapping," J. Mater. Chem. A 1, 9942–9946 (2013).

[8] Y. Gao, H. Cansizoglu, K. G. Polat, S. Ghandiparsi, A. Kaya, H. H. Mamtaz, A. S. Mayet, Y. Wang, X. Zhang, T. Yamada, E. P. Devine, A. F. Elrefaie, S.-Y. Wang, and M. S. Islam, "Photon trapping microstructures enable high-speed high efficiency silicon photodiodes," Nat. Photon. 11, 301–308 (2017).

[9] J. Gou, H. Cansizoglu, C. Bartolo-Perez, S. Ghandiparsi, A. S. Mayet, H. Rabiee-Golgir, Y. Gao, J. Wang, T. Yamada, E. P. Devine, A. F. Elrefaie, S.-Y. Wang, and M. S. Islam, "Rigorous coupled-wave analysis of absorption enhancement in vertically illuminated silicon photodiodes with photon-trapping hole arrays," Nanophotonics, 8(10), 1747-1756 (2019).

IOP Conf. Series: Materials Science and Engineering 1186 (2021) 012004 doi:10.1088/1757-899X/1186/1/012004

# Comparison of the antibacterial effect of a copper-coated surface on *Staphylococcus epidermidis* and *Pseudomonas putida* in different physiological states

**V Vibornijs[1], J Liepins[1], T Selga[2], V Bankovskis[3], P Cosemans[4], O Muter[1]**

[1]Institute of Microbiology & Biotechnology, University of Latvia, 1 Jelgavas Str., Riga LV-1004, Latvia

[2] Faculty of Medicine, University of Latvia, 1 Jelgavas Str., Riga LV-1004, Latvia

[3] Biosan Ltd., *7/2 Ratsupites Str., Riga, LV-1067, Latvia*

[4] SIRRIS, Wetenschapspark 3, B-3590 Diepenbeek, Belgium

olga.mutere@lu.lv

**Abstract**. The antimicrobial effect of Cu is dependent on several factors, including the physiological state of test-cultures. The present study aimed to compare the antibacterial effect of the newly developed Cu-coated surface, using bacterial cultures of *Staphylococcus epidermidis* and *Pseudomonas putida* harvested at exponential (6h) and stationary (24h) phases of growth as well as from a colony. Growth kinetics and cell morphology were characterised for each culture. The colony forming units log reduction values for a Cu-coated surface for cultures at exponential, stationary phase, and from a colony were 1.08, 2.85, and 2.14, respectively, for *S. epidermidis*, while 1.57, 3.95, and 3.14, respectively, for *P. putida*. The Cu-coated surface reduced the signal of ATP bioluminescence in both *S. epidermidis* and *P. putida*, with the more pronounced inhibition being for the cells at exponential and stationary phase of growth. The data indicated that bacterial cultures at the early exponential stage of growth are more resistant towards antimicrobial surfaces than cells at the stationary phase of growth and from a colony. This fact is of great importance for estimating the efficiency of antimicrobial coatings.

## 1. Introduction

Copper's antimicrobial properties have been widely applied as antimicrobial materials or coatings. Many of these products claim to kill 99.9% of disease-causing bacteria within two hours [1-2]. Cu ions reduce the potential difference on the bacterial cell membrane via binding to negatively charged domains, causing membrane depolarisation, leakiness, or even rupture [3]. Furthermore, the antimicrobial effect of Cu can be based on the reactive oxygen species (ROS) generation [2,4].

The antimicrobial effect of Cu is dependent on several factors, i.e., the physical form of Cu (ions or nanoparticles (NPs)), chemical state of Cu (oxidation state), and Cu concentration. Besides, the outcome of testing is largely dependent on the form of application, i.e., humidity, temperature, and composition of the chemical environment [5]. Furthermore, the physiological status of the tested culture, i.e., growth rate, subpopulations including antibiotic refractory, persisting, planktonic or immobilised cells, the density of the microbial population, and concentrations of exoproducts released into the medium either by living or dead cells, will influence the efficiency of an antimicrobial agent [6-7]. Thus, the outcome of the 'standard' *in vitro* antibacterial test can be misleading in terms of cell resistance towards antimicrobial surfaces.

We hypothesised that bacterial cultures obtained at the early exponential and stationary phase of growth, as well as from a colony, will demonstrate considerable differences in terms of resistance to antibacterial surfaces. These differences are expected to correlate with ATP synthesis, being the indicator of cell metabolic activity and resistance to oxidative stress.

Two strains representing Gram-positive (*Staphylococcus epidermidis*) and Gram-negative (*Pseudomonas putida*) bacteria were used in this study. Both cultures are widely used for testing antibacterial agents [8-9], and particularly, Cu nanostructures [10].

The aim of the present study was to compare the antibacterial effect of the newly developed Cu-coated surface, using bacterial cultures in different physiological states. Test cultures of *S. epidermidis* and *P. putida* were obtained at the early exponential (6 h) and stationary (24 h) stages of growth and from a colony. The testing was performed using the ISO 22196 / JIS Z 2801 protocol [11], with minor modifications.

## 2. Materials and Methods

### 2.1. Materials and cultures

Two bacterial cultures, i.e., Gram-positive *Staphylococcus epidermidis* MSCL 333 and Gram-negative *Pseudomonas putida* MSCL 650, were used. Both cultures were received from the Microbial Strain Collection of Latvia (MSCL).

The copper-coated and uncoated inox foil samples were prepared by Dr. Patrick Cosemans in cooperation with SIRRIS, Belgium. The Cu-coating was deposited by physical vapour deposition onto an inox foil using the Hauzer Flexicoat 850 equipment. Cu was deposited using magnetron sputtering at 200 °C.

### 2.2. Cultivation

*S. epidermidis* and *P. putida* were cultivated in a RTS-1 bioreactor (Biosan Ltd., Latvia), which operates based on the patented Reverse-Spin® technology. The optical density of the culture was automatically measured at 850 nm every 20 min, using a near-infrared optical system. The cultures were grown aerobically in 50 mL Falcon® tubes with 10 mL of Luria-Bertani (LB) broth, an agitation speed of 250 rpm, and a reverse spin interval of 2 s at 30 °C and 37 °C for choosing the optimum temperature conditions. The composition of the LB broth was as follows, g/L: tryptone, 10.0; yeast extract, 5.0; and NaCl, 10.0. The inoculum was used from 24 h bacterial cultures (100 µL/10 mL). Each experiment was performed at least in triplicate. Biomass was harvested after 6 h and 24 h, washed by centrifugation at 10,000 rpm for 5 min, and resuspended in 0.85% NaCl to a final cell concentration of $10^7$ CFU/mL. Additionally, both cultures were cultivated on the LB agar for 24 h. Biomass from a colony was resuspended in 0.85% NaCl. $OD_{620}$ measurements were performed on a Tecan microplate reader F50 (Switzerland).

### 2.3. Growth kinetics

The Kmax was calculated by plotting $OD_{850}$/time data to a ln($OD_{850}$)/time graph. The maximum liner slope of ln($OD_{850}$)/time curve is equal to the maximal growth rate and denoted as Kmax(1/h). The doubling time corresponded to ln2/K. Other data outside the $K_{max}$ site of the ln($OD_{850}$)/time curve represented the growth rate and was named K(1/h) (Equation 1).

$$K(1/hr)=\ln((OD_{850}(n))/OD_{850}(n-1))/(t(n)-t(n-1)) \quad (1),$$

where K: growth rate; (n) and (n-1): time interval.

### 2.4. Antimicrobial surface testing

Testing was performed according to [11] with minor modifications. Tested foil pieces were cut into 1 $cm^2$ regular squares using a scalpel. Samples were treated with 96% ethanol and additionally sterilised shortly in an open flame. Each test piece was placed in a cup of a sterile 15 mL tube; then, 10 μL of bacterial suspensions ($10^7$ CFU/mL) were placed on the testing surface and with sterile parafilm 0.64 $cm^2$ (parafilm was previously immersed in ethanol for 10 min, dried, and exposed to UV for 30 min). Tubes with test pieces were closed and left upside down at 23 °C for 60 min. All variants were prepared in triplicate. Afterwards, 200 μL sterile saline solution was added to the tube cup with the test piece, closed, and treated by ultrasound (US) in the US bath (50W) for 3 min. Cell suspensions (20 μL) from the tubes were immediately transferred to a microplate with 180 μL sterile deionised water for preparing the decimal dilutions. Samples were plated by a microdilution droplet method on the LB agar. The number of CFU was counted after a 24 h incubation at 30 °C. The conditions of test effectiveness were judged by preparing an additional sample with the untreated test piece and test culture, which was immediately plated without incubation.

The value of antimicrobial activity (LOG reduction) was calculated according to the following equation:

$$R = [\log (B/A)] - [\log (B/C)] \quad (2),$$

where R: value of antimicrobial activity; A: average of the CFU number immediately after inoculation on the untreated test piece; B: average of the CFU number on the untreated test piece after incubation; and C: average of the CFU number on the antimicrobial test piece after incubation.

### *2.5. ATP concentration*

Test pieces for ATP testing were prepared as described above. Incubation was performed in a 12-well microplate. After incubation, the ATP was extracted from the samples by adding 20 μL/well of a 10% trichloroacetic acid (TCA) mixed with an acetate base broth at a 1:1 ratio. The mixture was stirred for 30 seconds. Afterwards, the extraction was stopped by adding 120 μL acetate buffer (pH 7.75, 0.1 M). Five microlitres from each well were transferred to a new 96-well microplate, and 45 μL acetate buffer were added. Measurement process: The ATP Bioluminescence Assay Kit CLS II (Sigma-Aldrich) was used to quantify ATP in the sample. Ten microlitres of double distilled water were added to the reaction mix, and 1 mL 10% TCA was added to the ATP standard vial. The reaction mix was kept on the ice. Each sample was tested in triplicate. ATP standard dilutions were prepared in the same way as the samples were prepared. The ATP calibration graph was constructed for each batch of samples. The bioluminescence was recorded on the spectrofluorometer TECAN Infinite M200 (Switzerland). To start the reaction, 45 μL of the reaction mix was automatically added to the 50 μL sample, followed by immediate measurement of bioluminescence. Data were collected in Excel spreadsheets, and all further calculations were performed there.

### *2.6. Microscopy*

Bacterial cells were fixed with 96% ethanol and stained with propidium iodide (PI) at a concentration of 20 μmol/mL. The samples were analysed with a Leica DM RA-2 confocal laser scanning microscope (Germany) equipped with a TCS-SL confocal scanning head. The dye (PI) molecules were excited with a 488 nm laser band, and the fluorescence was detected between 600 nm and 640 nm wavelengths. The size of cells was measured as an average of >100,000 cells.

### *2.7. Statistical analysis*

The cultivation and antimicrobial testing were performed in triplicate. The microbiological and biochemical measurements were also done in triplicate. Thus nine data points were obtained for each experimental variant. The data presented in the figures are expressed as the mean value ± standard deviation. The differences between the treatments were assessed by the Student's t-test and one-way analysis of variance (one-way ANOVA) in Microsoft Excel, Office365.

## 3. Results and Discussion

### *3.1. Growth kinetics of S. epidermidis and P. putida*

Preliminary experiments showed that *S. epidermidis* grew faster at 37 °C, while *P. putida* grew faster at 30 °C in LB broth (data not shown). These results corroborated with data published by other authors [12-13]. Therefore, the biomass of *S. epidermidis* and *P. putida* for further antibacterial testing was obtained by cultivation at 37 °C and 30 °C, respectively.

The growth curve shapes of the tested bacterial cultures differed; namely, *S.epidermidis* cultures entered the stationary phase at the 13th hour of cultivation, while *P. putida* continued growing up to the 24th hour (Figure 1A). After 24 h of cultivation the $OD_{850}$ of *S.epidermidis* culture achieved 2.78 ($5.6*10^7$ CFU/mL), while *P. putida* was 1.26 ($1.2*10^7$ CFU/mL) (Figure 1A). Nevertheless, a decreasing growth rate (K value) was detected for both cultures at the 11th hour of cultivation, and it was 0.23 $\Delta OD_{850}/\Delta t$ and 0.16 $\Delta OD_{850}/\Delta t$ for *S. epidermidis* and *P. putida*, respectively (Figure 1B). Searching for the point on the growth curve where the cells start to proliferate, the 6th hour of cultivation was detected as the start of exponential growth for both *S. epidermidis* and *P. putida*, with the growth rate of 0.05 $\Delta OD_{850}/\Delta t$ (Kmax = 0.475) and 0.02 $\Delta OD_{850}/\Delta t$ (Kmax = 0.930), respectively (Fig. 1B). According to the $\ln(OD_{850})$/time curve, after 24 h of cultivation, the growth rate considerably decreased for *S. epidermidis* (K = 0.034) and *P. putida* (K = 0.001). So, for all further testing in this study, two bacterial cultures with contrasting physiological status (6 h – early exponential, 24 h – stationary phase of growth) were compared.

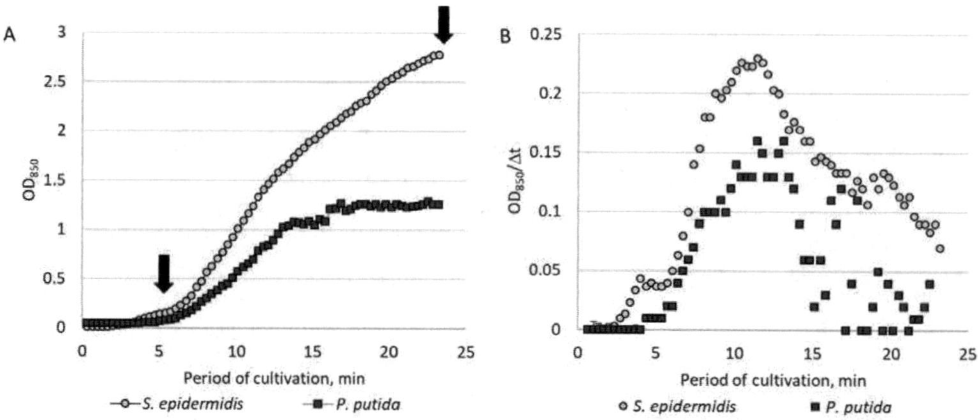


**Figure 1.** Growth kinetics of *Staphylococcus epidermidis* LMKK333 and *Pseudomonas putida* LMKK650. A: growth curve; B: growth rate. *S. epidermidis* and *P. putida* were cultivated in RTS-1 bioreactors at 250 rpm and 37 °C and 30 °C, respectively. Arrows indicated the sampling points of cultures at early exponential (6 h) and stationary (24 h) phases used for testing.

*3.2. Morphology of bacterial cells*

Visualisation of the tested bacterial cultures showed a species-specific response to the differences between growth stages (i.e., 6 h vs 24 h cultivation) and type of cultivation (i.e., broth vs agar). Specifically, spherical-shaped *S. epidermidis* cells did not differ in their size and remained in the range of 0.4-0.6 μm in the samples with 6 h and 24 h cultures and from the colony (Figure 2A, B, C). Instead, the cell morphology of *P. putida* cultures revealed considerable differences. The average *P. putida* cell length in 6h and 24h cultures and from a colony was 2-6 μm, 2-6 μm (43%) and 12 μm (57%), and 2-6 μm, respectively. The width of rod-shaped cells remained unchanged in all three cultures of *P. putida* and was 0.4-0.6 μm (Figure 2D, E, F).

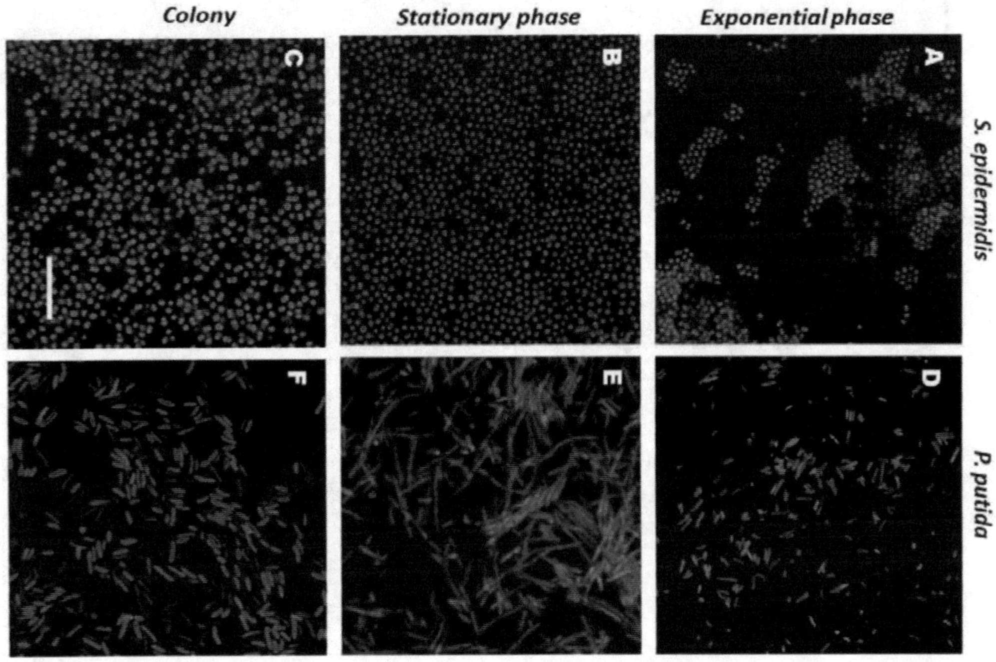


**Figure 2.** Micrographs of the *S.epidermidis* (A, B, C) and *P. putida* (D, E, F) cultures grown in LB broth till the early exponential phase (6 h) and stationary phase of growth (24 h), and on LB agar for 24h. Different colours on the cells correspond to the different laser scanning depths made by a confocal laser scanning microscope. Bar = 7.5 μm.

*3.3. Test for antimicrobial activity*

The log reduction values for a Cu-coated surface are shown in Figure 3. For *S. epidermidis* cultures, the log reduction for cells at exponential and stationary phase and from a colony was 1.08, 2.85, and 2.14, respectively; and for P. *putida* cells, it was 1.57, 3.95, and 3.14, respectively (Figure 3). A comparison of the results obtained for the two bacterial species revealed a higher resistance of *S. epidermidis* than *P. putida* to a Cu-NPs coated surface. Our data are in a good agreement with those recently reported by other authors, i.e. that Gram-positive bacteria tended to show a higher resistance to Cu-derived microbiocides than Gram-negative bacteria [14-16].

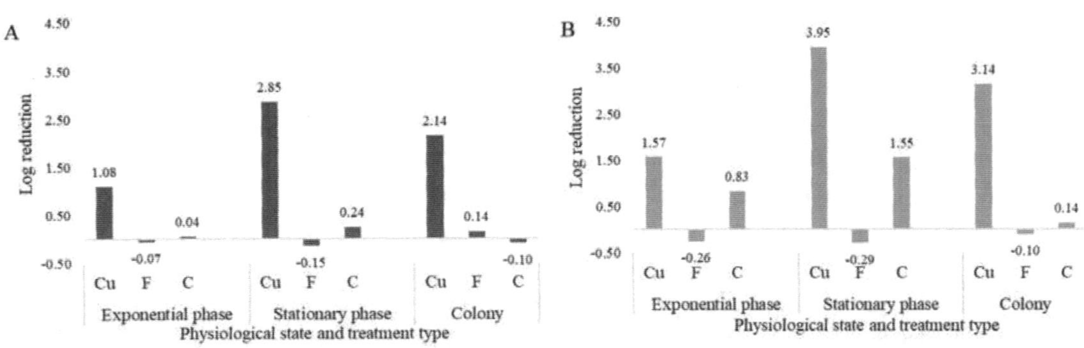


**Figure 3.** Log reduction of the bacterial cultures after 60 min contact with a Cu-NPs coating and respective controls. A: *Staphylococcus epidermidis*; B: *Pseudomonas putida*. Cu, F, and C: stainless foil with Cu-coating, non-coated foil, and control, respectively.

### *3.4. ATP snapshot data*

The concentration of ATP in bacterial cultures after treatment is shown in Figure 4. The Cu-coated surface reduced the signal of ATP bioluminescence in both cultures, with the more pronounced inhibition found in cells at the exponential and stationary phases of growth (Figure 4).

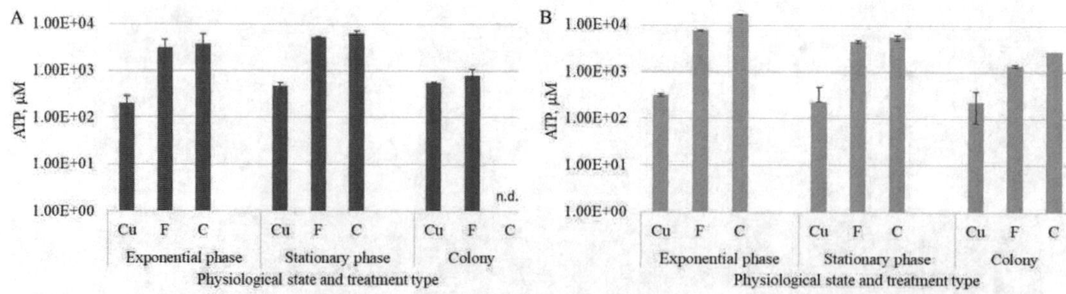


**Figure 4.** ATP concentration in bacterial cultures after 60 min contact with a Cu-NPs antimicrobial coating. A: *S. epidermidis*; B: *P. putida*. Cu, F, and C: stainless foil with coating, non-coated foil, and control, respectively.

The occurrence of ATP on the surface is a marker of the presence of microorganisms. ATP bioluminescence is a traditional and immediate test to detect microbial contamination or cleaning efficiency of a surface [17]. Even though it is a popular and quick method that is used in many contexts (as a fast method to detect contamination in the food industry, hospital inspections, etc.), this method also has some shortcomings. Depending on the method modification, there is a certain threshold of a minimal ATP concentration that can be detected and, subsequently, a minimal microbial CFU that can be detected by [18]. Therefore, while it is a fast and popular method for detecting contamination of the surfaces, due to its high signal threshold (several thousands of cells are necessary to measure ATP while few cells are enough to detect CFUs in colony growth assays), it is a rather qualitative tool for determining the log reduction of contaminant of the surface. For example, we found that Cu treated surfaces leave too little live CFUs to be detected by ATP luminescence, while colonies can still be detected using microbial growth assays after incubation with Cu treated surfaces.

Summarising the results obtained in this study, the following conclusions were made:

- *S. epidermidis* and *P. putida* cultures after 6 h of cultivation had the maximum growth rate, i.e., Kmax 0.475 and Kmax 0.930, respectively. This characterised cultures as exponentially grown and, thus, relevant to the comparative study on bacterial cultures with contrasting physiological conditions.
- Exponential *S. epidermidis* and *P. putida* cultures demonstrated the lowest CFU log reduction value, i.e., the highest resistance to the Cu-coated surface, compared to cells at the stationary phase of growth and from a colony.
- The signal of ATP bioluminescence on the Cu-coated surface was reduced in exponentially and stationary grown cells. We concluded that ATP measurements could speed up microbial testing of surface coatings if the minimal cell number varied between 3,000 and 15,000 CFUs for *S. epidermidis* and 8,000 and 9,000 CFUs for *P. putida*.

**Acknowledgements** The study was supported by COST Action 15114, AMICI (Anti-Microbial Coating Innovations to prevent infectious diseases).

**References**

[1] Grass G, Rensing C and Solioz M 2011 *Applied and Environmental Microbiology* **77**

1541–1547

[2] Mitra D, Kang ET and Neoh KG 2020 *ACS Applied Materials and Interfaces* **12** 21159–21182

[3] Fang L, Cai P, Chen W, Liang W, Hong Z and Huang Q 2009 *Colloids and Surfaces A Physicochem. Eng. Asp.* **347** 50–55

[4] Slavin YN, Asnis J, Häfeli, UO and Bach H 2017 *Journal of Nanobiotechnology* **15** 65

[5] Vincent M, Hartemann P and Engels-Deutsch M 2016 *International Journal of Hygiene and Environmental Health* **7** 585–591

[6] Wiegand C, Völpel A, Ewald A, Remesch M, Kuever J, Bauer J, Griesheim S, Hauser C, Thielmann J, Tonndorf-Martini S, Sigusch BW, Weisser J, Wyrwa R, Elsner P, Hipler UC, Roth M, Dewald C, Lüdecke-Beyer C and Bossert J 2018 *PLoS One* **13** e0194339

[7] Dalhoff A, Bowker K and MacGowan A 2020 *International Journal of Antimicrobial Agents* **55** 105809

[8] Wischer D, Schneider D, Poehlein A, Herrmann F, Oruc H, Meinhardt J, Wagner O, Ahmed R, Kharin S, Novikova N, Haag R, Daniel R and Grohmann E 2020 *Frontiers in Microbiology* **11** 1626

[9] Sorrentino E, Succi M, Tipaldi L, Pannella G, Maiuro L, Sturchio M, Coppola R and Tremonte P 2018 *International Journal of Food Microbiology* **266** 183–189

[10] Lacmanova V, Nguyenova HY, Ulbrich P, Slepicka P, Sajdl P, Svorcik V and Reznickova A 2020 *Materials Today Communications* **24** 101207

[11] Antimicrobial Products – Test for Antimicrobial Activity and Efficacy JIS Z 2801:2000. Japanese Standards Association, Tokyo, 2020.

[12] Noll P, Treinen C, Müller C, Senkalla S, Lilge L, Hausmann R and Henkel M 2019 *AMB Express* **9** 154

[13] Zou M and Liu D 2020 *Food Science and Human Wellness* **9** 370-376

[14] Kim T, Cho J, Cha D, Kim MS, Park EJ, Lee HJ and Lee C 2020 *Journal of Hazardous Materials* **400** 123305

[15] Roy S and Rhim JW 2020 *International Journal of Biological Macromolecules* **164** 37–44

[16] Videira-Quintela D, Guillén F, Montalvo G and Martin O 2020 *Colloids Surfaces B Biointerfaces* **195** 111216

[17] Omidbakhsh N, Ahmadpour F and Kenny N 2014 *PLoS One* **9** e99951

[18] Sanna T, Dallolio L, Raggi A, Mazzetti M, Lorusso G, Zanni A, Farruggia P and Leoni E 2018 *BMC Infection Diseases* **19** 583

# Assessment of the corrosion susceptibility of 434 ferritic stainless steel in chloride-sulphate solution

**Roland T. Loto**[1]

[1]Department of Mechanical Engineering, Covenant University, Ogun state, Nigeria

Email: tolulope.loto@covenantuniversity.edu.ng

**Abstract.** The corrosion polarization behaviour of 434 ferritic stainless (434ST) was studied in 1 M $H_2SO_4$ at 0.25% to 2% NaCl concentration through potentiodynamic polarization techniques and open circuit potential analysis. Results showed corrosion rate of 434ST at in 1 M $H_2SO_4$ solution at 0% NaCl concentration is 5.51 mm/y. However, at 0.25% NaCl concentration corrosion output decreased to 1.31 mm/y due to competitive adsorption mechanism with dissolved $O_2$ atoms. Increment in NaCl concentration causes increase in corrosion rate till 1% NaCl concentration at 5.97 mm/y. After 1% NaCl concentration, the corrosion of 434ST attained threshold deterioration mechanism with values varying between 6.20 mm/y and 6.56 mm/y. Shift in corrosion potential plot with respect to 434ST at 0% NaCl concentration indicates anodic corrosion and passivation mechanism on the steel surface which was also proven from the anodic Tafel slope values. Open circuit potential plots showed plots at 2% NaCl concentration showed the least thermodynamic tendency to corrode with the strongest shift to electropositive potentials which culminated at -0.439 V at 5400 s of observation. The plot at 0% NaCl solution exhibited the most electronegative plot configuration by reason of the deteriorating reaction of $SO_4^{2-}$ on the steel.

## 1. Introduction

Stainless steel is in the group of alloys composed of a minimum of 10.5% Cr content within their metallurgical structure. The presence of Cr in addition to Mb, Ni etc. imparts exceptional mechanical and corrosion resilience characteristics as against carbon steels [1, 2]. Cr in stainless steels reacts with $O_2$ in aqueous industrial environments to form a passive, inert $Cr_2O_3$ on the steel which protects it from the action of corrosive anions [3]. Different classes and grades of stainless steels exist worldwide tailored to specific applications such as those requiring structural strength, corrosion resistance, high temperature oxidation, microbial corrosion etc. However, inappropriate selection and application of stainless steels significantly impacts their operational lifespan especially in environments that degrade stainless steels. Deterioration of stainless steel components, structures and equipment are prevalent in environments consisting of sulphate, chloride, thiosulphate ions etc. by reason of reaction of the ions with the steel's exterior. This is common in petrochemical refineries, production of fertilizers, chemical vulcanization, energy production, marine applications and desalination plants. In addition to specific of industrial operating conditions, microstructural properties, important alloy composition and ionic concentration of corrosive species significantly influences the corrosion resilience of stainless steels [4–9] making the utilization of specific steels in some environment a huge economic waste [10]. Appropriate representation of the threshold corrosion resilience of stainless steels is important for their optimal application, sustainability of industrial architecture is of utmost importance. Generally, the corrosion resilience of stainless steels is by reason of the durability and electrochemical equilibrium of the protective oxide on their exterior [11–17]. The manuscript evaluates the threshold corrosion resistance of 434 ferritic stainless steel in 1 M $H_2SO_4$ electrolyte at 0.25-2% NaCl concentration. 434 stainless steel is a low carbon ferritic stainless steels with optimal corrosion resistance in mild aqueous corrosive conditions. The steel exhibits sufficient oxidation resistance at high temperatures. 434 stainless steels are ductile, not readily work hardened and can easily be fabricated.

## 2. Material and methods

The elemental constituents (wt. %) of 434 ferritic stainless steel (434ST) analysed with PhenomWorld scanning electron microscope as shown in Table 1. 434ST plates gotten from an automobile door trim was cut to 8 separate samples with dimensions of 1 $cm^2$. The 434ST were subsequently encased in vercocit resin mounts. The steels exterior was subjected to metallographic processing with SiC papers with 80, 320, 600, 800 and 1000 grits. They were subsequently burnished with 6 μm diamond burnishing fluid, and washed with purified $H_2O$ and $C_3H_6O$. 200mL of 1 M $H_2SO_4$ was concocted in 8 different sets from analar grade $H_2SO_4$ solution (98%). NaCl was added to the 1 M $H_2SO_4$ solution at 0%, 0.25%, 0.5%, 0.75%, 1%, 1.25%, 1.5%. 1.75% and 2% NaCl concentration. Potentiodynamic polarization measurement and open circuit potential analysis was done with a tripartite wire cell system inserted within a translucent container holding 200 mL of the $H_2SO_4$-NaCl solution and coupled to Digi-Ivy 2311 potentiostat. Potentiodynamic polarization graphs were drawn at sweep rate of 0.0015 V/s from -0.75 V - +1.75 V. Corrosion current density, $C_D$ (A/$cm^2$) and corrosion potential, $C_P$ (V) values were gotten from Tafel extrapolation of plots sections. Corrosion rate, $C_R$ (mm/y) was computed from the following numerical equation;

$$C_R = \frac{0.00327 \times C_D \times E_Q}{D} \quad (1)$$

$E_Q$ illustrates equivalent weight (g) of 434ST, 0.00327 stands for corrosion rate constant and $D$ illustrates density (g). Open circuit potential analysis of 434ST was done at step potential of 0.1V/s for 5400 s with dual-electrode cell system (Ag/AgCl standard electrode and 434ST electrode) coupled to Digi-Ivy 2311 potentiostatic instrument.

**Table 1.** Content (wt. %) of 434ST

| **434ST** | | | | | | | | | |
|---|---|---|---|---|---|---|---|---|---|
| Element | P | S | C | Cr | Ni | Cu | Mn | Si | Fe |
| % Content | 0.038 | 0.027 | 0.12 | 16 | 0.46 | 0.5 | 0.8 | 0.8 | 81.26 |

## 3. Results and discussion

### *3.1 Potentiodynamic polarization*

The corrosion polarization graphs of 434ST in 1M $H_2SO_4$ at 0% to 2% NaCl concentration are illustrated in Fig. 1. Fig. 2 depicts the magnified portion of the anodic-cathodic portion of the polarization graphs. Table 2 illustrates the values gotten from the polarization test. The reaction of $SO_4^{2-}$ and $Cl^-$ ions in the acid electrolyte significantly influenced the electrochemical behaviour of 434ST. 434ST corrosion polarization plots exhibited anodic deviation in corrosion potential in the acid electrolyte at 0.25% to 2% $Cl^-$ ion concentration relative to the plot from the electrolyte at 0% $Cl^-$ ion concentration (1 M $H_2SO_4$ solution). At 0% $Cl^-$ ion concentration, the corrosion potential of 434ST polarization plot is -0.462 V. However, the corrosion potential beyond this concentration varied between -0.434 V and -0.450 V. The anodic shift is by reason of the competitive reaction of $SO_4^{2-}$ and $Cl^-$ ions with the steel surface. The reaction of $Cl^-$ ions in the electrolyte induces a chloride-sulphate complex on the steel surface whose effect is deleterious and at best formation of a pseudo passive oxide. Nevertheless, evolution of a passive inert oxide on the steel surface occurs through adsorption and chemical combination with dissolved $O_2$ in the electrolyte. The $SO_4^{2-}$ and $Cl^-$ ions in the electrolyte reacts with the passive oxide and engages in competitive adsorption reaction with the dissolved $O_2$ on the steel surface resulting to a certain degree

IOP Conf. Series: Materials Science and Engineering 1186 (2021) 012005 doi:10.1088/1757-899X/1186/1/012005

in chloride penetration onto the substrate Fe metal due to its smaller size. This occurrence is excercebated when there are breakages in the oxide film as a result of the possible presence of flaws within the metallurgical structure of the steel. The flaws create discontinuities that allow $Cl^-$ ions and to a limited extent $SO_4^{2-}$ to diffuse and reacts with $Fe^{2+}$ causing its discharge into the electrolyte. The rate of discharge determines the degree of deterioration (corrosion) of 434ST. Table 2 shows the corrosion rate of 434ST in relation to $Cl^-$ ion concentration. At 0% $Cl^-$ ion concentration, corrosion rate of 434ST is 5.51 mm/y analogous to corrosion current density of 5.07E-04 A/cm$^2$. The corrosion rate value decreased to 1.31 mm/y at 0.25% $Cl^-$ ion concentration i.e. decrease in corrosion rate of 434ST at low $Cl^-$ ion concentration. However, increment in $Cl^-$ ion concentration induces increase in corrosion rate of 434ST till 1% $Cl^-$ ion concentration where the corrosion rate value is comparable to the value at 0% $Cl^-$ ion concentration. Beyond 1% $Cl^-$ concentration the corrosion rate values varied between 6.20 mm/y and 6.56 mm/y (1.25% to 2% $Cl^-$ ion concentration) conveying minimal $Cl^-$ ion concentration for increased deterioration of 434ST. The cathodic Tafel plot values were generally stable compared to the anodic counterparts due to stability of the $H_2$ evolution reactions which was less dominant compared to anodic dissolution reactions which varied significantly with respect to $Cl^-$ concentration. Generally, the corrosion rate results show $SO_4^{2-}$ ions with the presence of chlorides are deleterious to the thermodynamic stability of 434ST due to it relatively higher corrosion rate compared to 434ST at low chloride concentrations. Secondly, chloride ions have a threshold concentration were there electrochemical action has curbed effect on the corrosion resilience of 434ST. Thirdly in the presence of low chloride concentrations opposing adsorption between $SO_4^{2-}$ and $Cl^-$ ions decreases the thermodynamic inclination of 434ST to corrode.

**Table 2.** Potentiodynamic polarization values for 434ST corrosion in 1 M $H_2SO_4$ solution at 0% to 2% NaCl concentration

| Sample | NaCl Conc. (%) | Corrosion Rate (mm/y) | 434ST Corrosion Current (A) | Corrosion Current Density (A/cm2) | Corrosion Potential (V) | Polarization Resistance, $R_p$ (Ω) | Cathodic Tafel Slope, $B_c$ (V/dec) | Anodic Tafel Slope, $B_c$ (V/dec) |
|---|---|---|---|---|---|---|---|---|
| A | 0 | 5.51 | 1.01E-03 | 5.07E-04 | -0.462 | 25.37 | -6.611 | 2.957 |
| B | 0.25 | 1.31 | 2.40E-04 | 1.20E-04 | -0.450 | 68.69 | -7.264 | 2.246 |
| C | 0.5 | 3.10 | 5.71E-04 | 2.85E-04 | -0.434 | 45.03 | -7.888 | 4.306 |
| D | 0.75 | 3.84 | 7.07E-04 | 3.53E-04 | -0.436 | 40.37 | -8.082 | 6.468 |
| E | 1 | 5.97 | 1.10E-03 | 5.49E-04 | -0.448 | 22.37 | -7.111 | 5.519 |
| F | 1.25 | 6.56 | 1.21E-03 | 6.03E-04 | -0.446 | 16.30 | -7.327 | 6.468 |
| G | 1.5 | 6.20 | 1.14E-03 | 5.70E-04 | -0.439 | 18.53 | -8.200 | 5.109 |
| H | 1.75 | 6.25 | 1.15E-03 | 5.75E-04 | -0.442 | 18.48 | -7.781 | 6.351 |
| I | 2 | 6.36 | 1.17E-03 | 5.85E-04 | -0.443 | 19.49 | -8.609 | 7.615 |

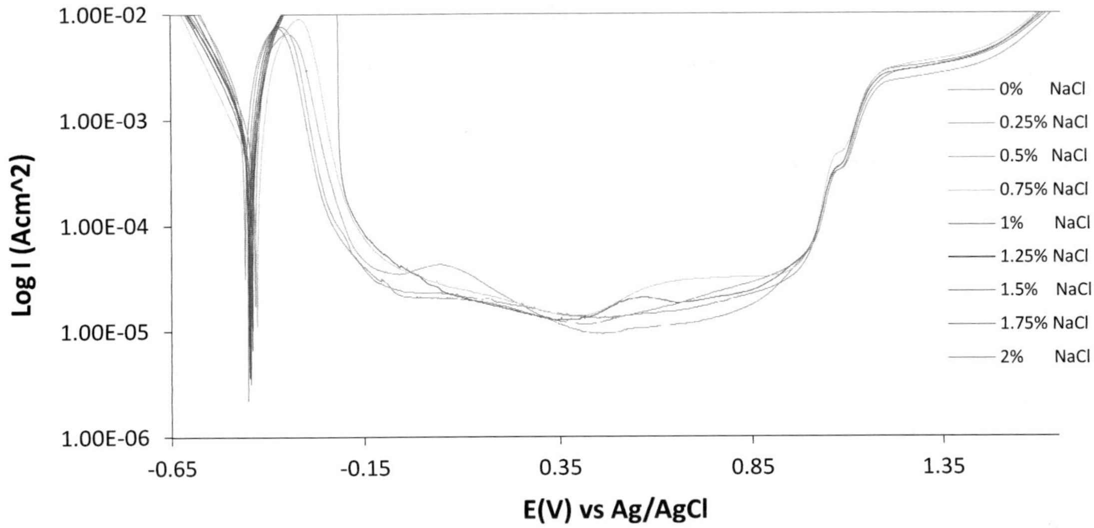


**Figure. 1.** Potentiodynamic polarization graphs for 434ST corrosion in 1 M $H_2SO_4$ solution at 0% to 2% NaCl concentration

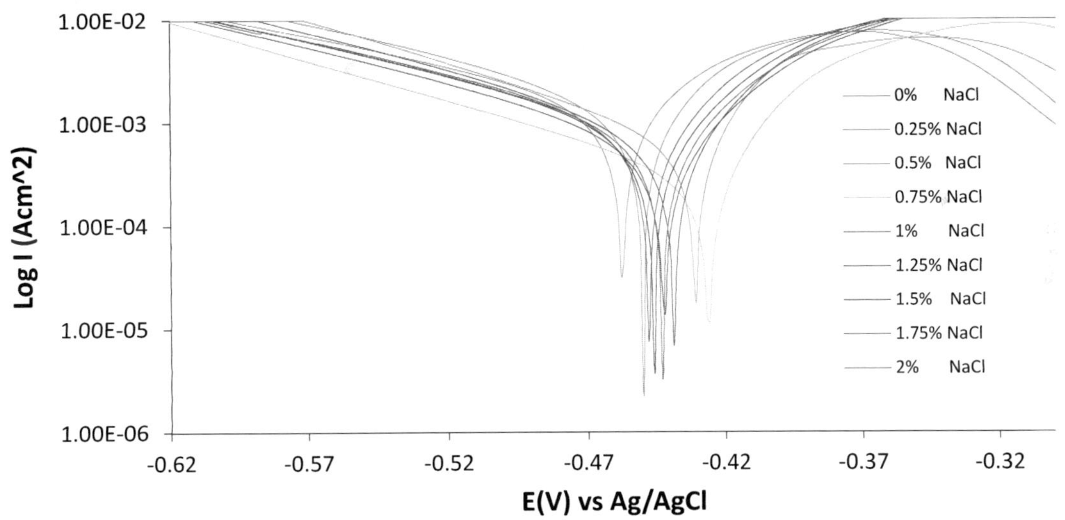


**Figure. 2.** Magnified view of the anodic-cathodic polarization plots for 434ST corrosion in 1 M $H_2SO_4$ solution at 0% to 2% NaCl concentration

*3.2 Open circuit potential analysis*

Open circuit potential graphs for 434ST in 1 M $H_2SO_4$ at 0%, 0.25% and 2% $Cl^-$ ion concentration are illustrated in Fig. 3. Generally, the potential plots at 2% $Cl^-$ ion concentration exhibited the most electropositive potential displacement while the plots at 0% $Cl^-$ ion concentration exhibited the most electronegative displacement. The plot at 2% $Cl^-$ ion concentration started at -0.442 V (0 s) and steeply deviated to electronegative potentials till -0.457 V at 300 s due to initial surface oxidation of the steel resulting from the synergistic reaction effect of $SO_4^{2-}$ and $Cl^-$ ions inside the electrolyte. This occurs when the protective inert oxide on the steel is yet to completely offer meaningful protection. Beyond 300 s of exposure the potential of the steel progressively shifts to electropositive potential as the protective oxide achieves relative stability and continues to grow. The potential plot at 0% $Cl^-$ ion concentration at -0.460 V (0 s) and shifted to -0.466 V at 107.70 s by reason of anodic reaction of the steel exterior in the vicinity of $SO_4^{2-}$ ions. However, the electronegative portion of the potential plot is significantly shorter than the one observed for the plot at 2% $Cl^-$ ion concentration. This is by reason of the relatively large size of $SO_4^{2-}$ ions compared to $Cl^-$ ions which can more easily diffuse through the protective film on the steel. As a result, evolution of the oxide occurred at much shorter time. Beyond -0.466 V, electropositive shift of the plots was visible till -0.448 V at 5400 s. The final plot configuration at 2% and 0% $Cl^-$ ion concentration shows the presence of $Cl^-$ ions with $SO_4^{2-}$ though debilitating increases the corrosion resistance of 434ST compared to $SO_4^{2-}$ ions only. The potential plot at 0.25% Cl- ion concentration displayed near similar plot configuration with the plot at 0% $Cl^-$ ion concentration, though its electronegative potential portion of the plot is visibly shorter and the final potential at 5400 s is more electropositive at -0.446 V.

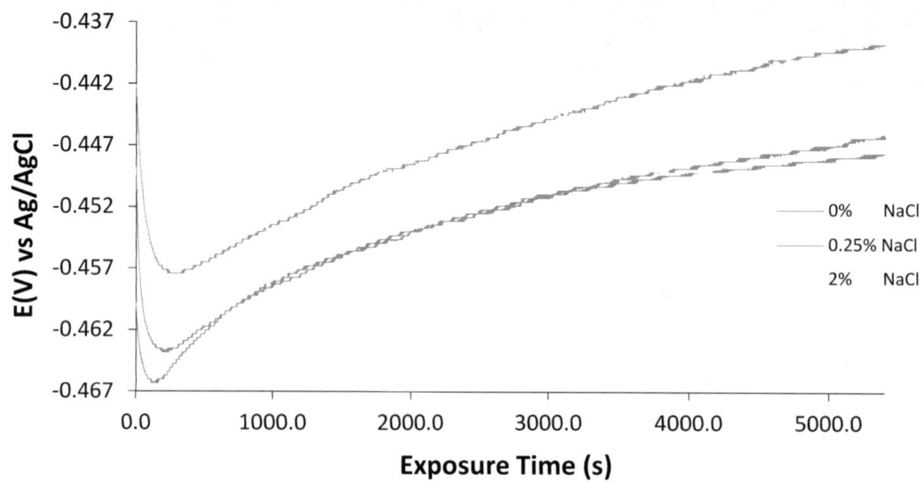


**Figure. 3.** Open circuit potential graphs of 434ST corrosion in 1 M $H_2SO_4$ solution at 0%, 0.25% and 2% $Cl^-$ ion concentration

## 4. Conclusion

434 ferritic stainless steel exhibited limited resistance to corrosion in the presence of $SO_4^{2-}$ ions. In the face of low concentrations of $Cl^-$ ions, corrosion of 434 steel decreased significantly due to competitive adsorption and reaction processes with the steel. Increment in $Cl^-$ ion concentration caused progressive increase in corrosion rate till a threshold chloride value were the corrosion rate was relatively high compared to the solution in absence of $Cl^-$ ions. Polarization plots showed significant anodic shift by reason of evolution of protective inert oxide on the steel. Open circuit potential plots depict the presence of $Cl^-$ ions increases the passive protective on the steel hence its corrosion resilience compared to the plot in the absence of chlorides.

## Acknowledgement

The author declares the support of Covenant University Ota, Ogun State, Nigeria for the execution of this research.

## References

[1] Black, J. (2006). Biological performance of materials: Fundamentals of Biocompatibility. 4th ed. Boca Raton: CRC Press.

[2] Čihal, V. (1999). Stainless steels and alloys. (in Czech). Praha: Academia.
[3] Incoloy Alloy 800H & 800HT, Technical Bulletin, Special Metals Publication SMC- 047. http://www.specialmetals.com [Retrieved 1st April, 2019].
[4] Liptáková, T. (2011). Pitting corrosion of stainless steels. Materials Engineering, 18, 115–20.
[5] Szklarska Salowska, Z. (2005). Pitting and crevice corrosion. Houston Texas: NACE International.
[6] Rodabaugh, R.D. (1999). Surface cleaning. ASM Handbook. 3rd ed. USA: ASM International.
[7] Feng, H., Li, H., Wu, X., Jiang, Z., Zhao, Si., & Zhang T, et al. (2018). Effect of nitrogen on corrosion behaviour of a novel high nitrogen medium-entropy alloy CrCoNiN manufactured by pressurized metallurgy, Journal of Materials Science and Technology, 34(10), 1781–1790.
[8] Burstein, G.T., & Pistorious, P.C. (1995). Surface roughness and the metastable pitting of stainless steel in chloride solutions, Corrosion, 51(5), 380–385.
[9] Zhang, S., Li, H., Jiang, Z., Zhang, B., Li, Z., & Wu, J. et al. (2019). Effects of Cr and Mo on precipitation behavior and associated intergranular corrosion susceptibility of superaustenitic stainless steel S32654, Materials Characterization, 152, 141–150.
[10] Burkert, A., Lehmann, J., Burkert, A., Mietz, J., & Gumpel, P. (2013). Technical and economical stainless steel alternatives for civil engineering applications, Materials and Corrosion, 64, 1–17.
[11] Lee, J.B. (2006). Effects of alloying elements, Cr, Mo and N on repassivation characteristics of stainless steels using the abrading electrode technique, Materials Chemistry and Physics, 99(2–3), 224–234.
[12] Loto, R.T., & Loto, C.A. (2019). Evaluation of the localized corrosion resistance of 316 L austenitic and 430Ti ferritic stainless steel in aqueous chloride/sulphate media for application in petrochemical crude distillation units, Materials Research Express, 6, 086516. https://doi.org/10.1088/2053-1591/ab1a11.
[13] Loto, R.T. (2018). Potentiodynamic polarization studies of the pitting corrosion resistance and passivation behavior of P4 low carbon mold steel in chloride and acid chloride solution, Material Research Express, 5, 116509https://doi.org/10.1088/2053-1591/ aadc84.
[14] Loto, R.T. (2018). Effect of elevated temperature on the corrosion polarization of NO7718 and NO7208 nickel alloys in hot acid chloride solution, Journal of Bio and Tribo Corrosion, 4(71). https://doi.org/10.1007/s40735-018-0190-8.
[15] Loto, R.T., & Loto, C.A. (2017). Potentiodynamic polarization behavior and pitting corrosion analysis of 2101 duplex and 301 austenitic stainless steel in sulfuric acid concentrations, Journal of Failure Analysis and Prevention, 17(4), 672–679. https://doi.org/10.1007/s11668- 017-0291-6.
[16] Loto, R.T. (2017). Study of the corrosion resistance of type 304L and 316 austenitic stainless steels in acid chloride solution, Oriental Journal of Chemistry, 33(3), 1090–1096. https://doi.org/ 10.13005/ojc/330304.
[17] Loto, R.T. (2017). Electrochemical corrosion characteristics of 439 ferritic, 301 austenitic, S32101 duplex and 420 martensitic stainless steel in sulfuric acid/NaCl solution, Journal of Bio and Tribo Corrosion, 3(24). https://doi.org/10.1007/s40735-017-0084-1

# AUTHOR INDEX

**Institute of Physics**
Dirac House, Temple Back
Bristol BS1 6BE UK

ISSN: 1757-8981
ISBN 978-1-7138-4646-8